SEND ME

Chronicles of an FBI Sniper

Special Agent (Ret.)

Jeremy D.O. Rebmann

Copyright © 2025 by Jeremy Rebmann

All rights reserved.

No part of this book may be reproduced, stored in a retrieval system or transmitted in any form or by any means, electronic, mechanical, photocopying, recording, or otherwise, without express written permission of the author.

ISBN 979-8-9894716-0-7 (paperback)

ISBN 979-8-9894716-2-1 (digital)

ISBN 979-8-9894716-3-8 (hardback)

First Edition

Printed in the United States of America

Scripture quotations are from the ESV® Bible (The Holy Bible, English Standard Version ®), copyright © 2001 by Crossway Bibles, a publishing ministry of Good News Publishers. Used by permission. All rights reserved.

The views and opinions expressed in this product are that of the author and not that of the FBI or any government agency. This product has been approved by the FBI Pre-Publication Review Office. Limited changes from the original manuscript were required by the FBI to protect identities, tactics, techniques, and procedures.

To God, who made me, loves me, forgave me, saved me, and continues to save me every day.

To my bride, who put up with me leaving in the middle of the night, not knowing when I would be home.

To my daughter, the apple of my eye. My bud. Being your Dad makes me feel capable of heroism.

To Mom, who raised me and taught me to love what is right and noble.

To Joy, for convincing me that these stories need to be told.

To the men of FBI SWAT for a million laughs, for their incomparable bravery, and for welcoming me into the fellowship of the brotherhood.

AUTHOR'S NOTE

    These are the facts as I recall them. The missions are numbered as they appear in my journal. They are not all-inclusive, and therefore the mission numbers do not represent the total number of missions I had executed to date. I have conferred with my teammates to make sure these stories are told accurately.

    Most of the events described in this product unfolded on mornings that started at 2 a.m. and were fueled by caffeine, chaos, and adrenaline. These memories reside in my mind as a montage of events that blur together in a tapestry of excitement, exhaustion, joy, stress, and occasionally, fear. This is a glimpse into the life of a career FBI sniper.

TABLE OF CONTENTS

Foreword vi

Preface viii

1. New Guy 1
2. Carney Hostage Rescue 12
3. Glass Mountain 24
4. Operation Delta Blues 44
5. Wewoka 68
6. Rocket Bomber 87
7. DIGPRO 115
8. Sanctioned Killing 130
9. Operation Mischief Mayhem 145
10. Dirty Bomb 158
11. Cartel Safehouse 164
12. Dripping Springs 177
13. Operation Vatos Locos 200
14. Whom Shall I Send? 215

Glossary 222

Foreword

Dave Grossman

This book is the best cop book I have ever read! (And I have read a lot.) The author's amazing background, qualifications, and experience all work to bring this vital subject alive. Additionally, it is presented by a true warrior-wordsmith. The writing here is powerful and effective. And it is funny! Consistently, persistently, just danged funny. So many great chuckles in here. The kind of "cop humor" that puts you inside the mind and spirit of those who fight the good fight every day.

Most importantly, this is an amazing insight into the world of modern SWAT, FBI, Federal Agents, and police officers, and the criminals they live to apprehend. Future generations will look back in awe upon our magnificent "SWAT dogs" who rose to the challenge to defend our "flock" and "hunt the wolf" much as we think of the heroes and lawmen of the wild west. You see how very badly we need our mighty protectors and "sheepdogs" like my good friend, Special Agent Jeremy Rebmann, "for such a time as this."

In this book you will see that one of the greatest achievements of our SWAT teams is that "no one had to die today." As the author tells us, "We brought overwhelming force and numbers to a potential fight, and the usual result was that no one wanted to fight us. Surrender was the norm."

This book is a powerful insight as to how the elite of the elite operate. Every American knows about the SEALs and Delta Force, elite spec ops warriors fighting the good fight around the world. Now you have the best possible insight into the lives of those competent and compassionate warriors who perform the same mission here at home. Yes, "compassionate" might be the very best description for the SWAT operators you will meet in this book.

Empathy says, "What if this was my child?" Or, "What if it was my spouse who was murdered?" Empathy says, "What if it was my father, bleeding out his life and urgently needing medical care?" Empathy is what makes a great cop, a great SWAT operator, and a true sheepdog. Empathy is love in action. Empathy is love with its shoes on. These courageous operators fight evil with love. Evil is the absence of love, just as darkness is the absence of light. The author writes that they were "fueled by caffeine, chaos, and adrenaline." And, I add, motivated by love.

Jesus said, "Greater love has no one than this, that someone lay down his life for his friends (John 15:13)." So, what manner of love is this, that sheepdogs like Jeremy Rebmann, walk out of their homes, leave their beloved family every day, ready to lay down their lives for strangers…for people they have never met, until their moment of greatest need?

And I tell you that there are many ways to lay down your life. Sometimes the greatest love is not to sacrifice your life, but to live a life of sacrifice. And that applies to all of our first responders, doing a dangerous, dirty, and too often thankless job, putting their lives on the line every day, that others may live.

Here is the critical, essential difference between our military who often are ordered to kill, and our law enforcement who are always trying to save lives. Law enforcement officers use "deadly force" because they believe that there is no other option in the face of an immediate threat of death or grievous bodily harm. The moment this criminal is no longer a threat, that life is as precious as any other life on the planet, and the cops will strive to save that life.

When they do have to kill, the impact is very real.

Thus, I tell you that this book is the single best resource for anyone who wants to know what combat is really like, or how a SWAT team really operates. Anyone who enjoys true crime will love this book. Anyone who wants to know about the FBI, SWAT teams, or law enforcement operations needs to read this book. (And any "gear guys" out there will definitely like this book.)

Anyone who wants to get a powerful insight into the heroic law enforcement officers holding back the darkness. So, read, learn, enjoy, and buckle your seatbelt because it is an awesome ride!

Lt. Col. Dave Grossman, USA (ret.)
Author of *On Killing, On Combat, On Spiritual Combat,* and over a dozen other books.
www.GrossmanOnTruth.com

Preface

WE ONLY HAVE EACH OTHER

The engine revved in our US Air Force Ford Mustang as the driver slipped it into gear. Our front tires were touching the top edge of the numbers "32," painted on the concrete in giant lines of white paint. Each numeral was 80 feet across – big enough for someone to see from 5,000 feet up. Here at the end of Runway 32, we waited for our jet as it approached us from behind. It had been on a long glide from 70,000 feet and was down to 4,000 feet now.

My head pushed back into the headrest as the driver, Dan, a spy plane pilot himself, pressed hard into the accelerator pedal. We had three miles of runway ahead of us in which we would make our rendezvous with an exhausted pilot and his steed. He was low on fuel and ready to be home. The 5.0-liter V8 roared as we accelerated up to 100 miles per hour. Each time it shifted, I got a little buzz from the acceleration. I checked the stack of radios in the center console. We had UHF, VHF, and HF radios so that we could talk to our bird on all available frequencies. My flightline guys had triple-checked the crypto on these radios before we rolled out onto the flight line so we could transmit without local scanners hearing us.

I was an Air Force First Lieutenant on a special mission to maintain and launch US jets in a foreign country. We were instructed to wear civilian clothes and rent civilian cars from the local economy. We even shopped at the local grocery stores and lived in hotel rooms in town. No clothing with English lettering was allowed and even blue jeans and t-shirts were discouraged. It was hard not to stand out, though. I'm not sure we were fooling anyone. They knew we were Americans, but I don't think they would have guessed what we were doing.

This deployment seemed luxurious compared to the "tent cities" that we usually occupied for months at a time somewhere in the desert on the edge of a runway. In the tent cities, you didn't sleep much due to the afterburners shrieking across the runway at all hours of night. Our current off-base hotel and the rental cars were perks, but none of us spoke the language and we had no interpreters. It was an adventure to say the least. We were a couple months into this mission now, so I had gotten the hang of navigating my way to the base and changing into my uniform in my shop at the flight line. I was twenty-four years old, the keeper of six multimillion-dollar jets, and the commander of fifty troops. The Air Force was certainly a great place to get leadership experience fast.

"Tower, Pinyon Niner One One, short final, gear down and locked," the Lockheed U-2 pilot called out over the radio. His voice was unfamiliar to me, but he and Dan were squadron-mates. I was just the guy who made sure the jets got fixed, fueled, and launched on time.

I could just make out the outline of the sleek black plane as I craned my neck over my right shoulder. It looked almost like it was hovering, but it was going 100 miles per hour just like us. Dan eased up on the throttle to match the deceleration of the jet behind and above us.

The Mustang slowed to 90 miles per hour as the ominous shadow of the spy plane settled over us. Its enormous 100-foot wingspan covered half of the width of the giant runway. I looked out of the passenger window and saw the long, slim fuselage of our U-2, also known as the Dragon Lady, as it settled into ground-effect over the runway. The distinctive squeal of its turbojet engines eased up as the pilot spooled down to idle power. It was a giant aluminum glider now. It was graceful from my perspective, but on the inside the exhausted pilot was working frantically to keep the plane under control. Dan told me the Dragon Lady was notoriously difficult to land. It was hard enough that you needed another pilot to race down the runway in a sports car to talk you down.

"Ten feet. A little nose high. Boards out," Dan said over the radio. He had one hand on the wheel and another on the radio. He was staring diligently at the hulking jet next to us while we roared down the runway at highway speeds. In response, the pilot in the cockpit just fifty feet away deployed his spoilers to increase his deceleration.

We continued to match the speed of the spy plane as it settled down onto its two bicycle-configuration landing gear. When it took off eight hours ago, as designed, it had jettisoned its underwing landing gear, known as "pogos." Punching off the pogos saved weight and drag, and the remaining nose and tail gear retracted into the fuselage. When it came time to land, the pilot had to do the delicate balancing act of landing on just the bicycle gear.

We were in a perfect position under his left wing while the pilot struggled to keep the plane level and keep the wingtips from hitting the edge of the runway. The Mustang was decelerating in sync with the jet. The bicycle gear screeched and made a little black puff of burning rubber as the spy pilot delicately set his plane on the ground. As both the jet and the car eased to a stop, the pilot pushed his stick to the left and his left wingtip fell perfectly in place in front of our car. My flightline crew ran onto the runway to re-install the pogo gear.

"Hop on," Dan said to me with a smile. We had both graduated from the U.S. Air Force Academy, but he was a couple years ahead of me, so he was a Captain already. Little did I know that I would be at Air Force Special Operations Command HQ in Florida as an Air Force OSI Special Agent in less than two years, and I'd "pin on" Captain with an M11 9-mm concealed under my polo in a plain-clothes ceremony. Our careers had gone in different directions, but we had run into each other in this foreign country on a giant patch of runway.

As a Sortie Generation Flight Commander, I made the effort to attempt every job that my men did, and I hadn't done this particular one yet. The hot, dry air hit my face as I scrambled out of the Mustang. The summer heat made the runway a frying pan, and the burning JPTS jet fuel smelled like diesel exhaust from a truck stop. I pulled myself onto the top of the Dragon Lady's left wing. I positioned my head forward, feet back, hands holding onto the leading edge of the wing. One of the flightline crew climbed up and laid down next to me. According to the Technical Orders for this jet, she needed two personnel that were of combined sufficient weight to keep the wing pressed down while the opposite pogos were installed.

The wing was bitter cold, a stark contrast to the heat from the runway, and it caught me by surprise. My fire-proof Nomex flight gloves stuck to the leading edge of the wing like wet fingers on a dry ice cube. This big black spy plane had just returned to the ground from being on the verge of space.

With the pogos installed, I climbed back into the Mustang. The pilot throttled back up to make a graceful right turn off of the active runway and we followed it down the taxiway until we got back to the maintenance hangar. The canopy came up on the U-2. The pilot sat perfectly still in his space suit. At 70,000 feet your blood will boil, so they wore pressure suits like astronauts. I thought he would've jumped up to get out of that plane as soon as possible. After all, he had been cramped into the volume of a refrigerator, unable to use the bathroom or scratch his own nose in that pressure suit for eight hours.

There was a surprisingly large group of people waiting at the hangar. Some of them were off-duty and wearing civilian clothes. Some of them were pilots, some flightline crew, and some were admin. They were cheering as two people scrambled to help the exhausted pilot get up and carried him down the plane's staircase. I had spent most of my Air Force career thus far launching KC-135 Stratotankers and was used to seeing the aircrew leisurely climb out of the jet while telling stories before they boarded a blue air-conditioned crew bus that drove them to Base Ops. This pilot's astronaut helmet came off and I could finally see his weary face. They continued to cheer and took turns rubbing the top of his sweaty head as they carried him into the hanger and helped him get his pressure suit off. I walked over to the driver-side door of the Mustang and looked at Dan.

"What's going on here? Did he just do something special?" I asked.

Dan looked at me and said, "Of course he did something special. He went into enemy territory and came home alive."

I said, "You mean to tell me that you guys celebrate after every single mission?"

Dan's face was serious. "Absolutely. No one really understands what we do or the danger we face. We only have each other."

Those words echoed in my mind for the next twenty-six years as I served high-risk warrants and conducted protective details in the Air Force Office of Special Investigations (OSI) and in the FBI as a Special Agent and SWAT operator. Dan taught me that some warriors stay on the battlefield even when the rest of the world is in peacetime. I would understand him better years later when I belonged to my own group of warriors.

In 2023, soon after I retired, there were just over 1,600 high-risk FBI SWAT operations stateside, and the numbers have been on the rise. There were 1,800 operations in 2024. That's two or three high-risk operations per month, per team – almost 150 ops a month nationwide. These weren't just simple arrest warrants; these included kidnappings, top-ten fugitives, bomb-makers, counter-terrorism ops, hostage rescue, chemical weapons, and counter-narcotics missions against ruthless cartels. At the outset of my career, in the early 2000s, we conducted these risky operations when armored vehicles were not yet available and before we had enough ballistic plates to equip the entire team. No robots, no drones, no thermal images from air support. Just men facing danger head-on. But we knew we were never alone.

I wrote this book so you could see inside the band of protectors that I trusted with my life so many times. During my 23-year FBI career, I went into harm's way with my brothers and sisters on over 200 high risk missions. Sometimes I was by myself in the woods with my rifle, sometimes I was solo on surveillance, but I was never alone. There was no extra pay or reward for the job, and we rarely got recognition for the thousands of hours of intense training and dangerous operations we experienced. I finally understood the reason why U-2 aircrew intentionally celebrated each mission.

We were warriors in a peacetime battlefield. No one really understood what we did or the dangers we faced. We went into enemy territory and came home alive.

We only had each other.

Chapter 1

New Guy
Mission #2
May 2001

METH LAIR

"Two minutes out," DB's voice boomed inside our stripped-out panel van. The van plowed over a pothole and a stray piece of .45 brass jolted into the air. It hovered in front of my face for a split-second like a metallic hummingbird before crashing back into the corrugated floorboard. As the brass slammed into the floorboard, so did my knees. *Ugh! I know I should have worn my knee pads.* I needed to protect my knees, but the pads cut off the blood flow to my lower legs. *I can't believe I'm really doing this.*

There's no shortcut to becoming an FBI SWAT operator. The average new Special Agent graduates from the FBI Academy in Quantico at the age of thirty-one. I was the second youngest in my class at twenty-seven. The oldest was thirty-seven. Before that, I had spent four years at the US Air Force Academy, followed by three years launching jets, and three years as a Special Agent in the USAF Office of Special Investigations (OSI). OSI is a fantastic investigative agency, but FBI SWAT was a siren song for me – the FBI's tactical program is unmatched in federal law-enforcement. After almost five months at New Agent Training at Quantico, and after another eighteen months working as a street Agent, I was finally eligible to try out for SWAT. I was stunned when I was picked up on the first selection. Here I was, a "new kid" at twenty-nine years old, finally on the team.

It was early 2001, and this was only my second operation, but it was the first one in which I would be making entry into a residence filled with armed and dangerous drug dealers. My first mission had been uneventful. It involved me just standing on perimeter security behind a drug dealer's house while the senior guys cleared it. It was highly uncommon to let a new operator enter and clear a structure with the senior guys and I intended to make a good impression today.

The van was stripped out and, other than the front two seats, the rest of us were sitting or kneeling on the sheet-metal floor of the van. SWAT operators got better gear than most Agents, but it still wasn't exactly ideal. Clinton had recently finished his second term, the Cold War had fizzled out, and other than Desert Storm, there hadn't been much emphasis on military and police protective equipment since the Vietnam War.

The front panel of my body armor dug into my quads and was cutting off the blood flow. When JimBo, Z, PC, and I had passed selection, we were allowed to pick through whatever armor and gear was left over in the team room in the garage at HQ. There were no new equipment purchases for new operators back then. My armor was sized "Large/Extra Long" and, at 5' 9", I was not "extra long." It was so large that it hung down over my belt, so I had my pistol, extra pistol magazines, MP5 magazines, and flash-bangs on belt extenders strapped to the thighs of both legs.

Most of my teammates carried extremely accurate, custom Springfield 1911 .45-caliber pistols with ten-round extended magazines protruding from the grip. The decision to standardize firearms across all fifty-six field SWAT teams would not come until after the Branch Davidian stand-off in Waco where numerous teams arrived with various pistols and in various calibers. The .45 was now the standard pistol caliber across all FBI SWAT teams. I was too new to the team to be issued a custom 1911, so I had gotten permission to carry my reliable personally-owned .45-caliber Glock. I was glad to carry it instead of the .40-caliber Glock model 23 that was standard-issue for new Agents, because that model was unreliable when a weapon light was attached to it.

"One minute out!" DB called. DB was a former police officer and our Assistant Team Leader (ATL). I was intimidated by him. Most of us were. I had seen him tear into other junior operators for minor mistakes and I feared his wrath. He was wearing the same black BDU's

(Battle Dress Uniform) and black leather boots as I was, but his uniform had faded to a dark gray from all the abuse it had taken over the years. Mine was practically fresh out of the box and had that "new guy" look. We all wore woodland-camouflage BDUs for training and saved the black BDU's for operations. With black armor, black uniforms, black boots, and black helmets, the first thing the "bad guys" would see coming in the night was the white "FBI" placard on the front of our body armor.

DB must have liked me at least a little bit, because he opted to carry his M4 rifle for this op and he had lent me his full-auto 10mm MP5 submachine gun, usually only carried by the most senior operators. It had an adjustable stock and a flashlight built into the forend, as well as an Aimpoint red dot sight. It was sexy, powerful, state of the art, and always turned heads. My assigned MP5 was a simple semi-automatic 9mm with a fixed stock and basic iron sights. It was essentially a big 9mm pistol. The best part about DB's weapon was the built-in flashlight. Mine didn't have one. Clearing a dark crime lair full of armed drug dealers with no light on my weapon would have been an unpleasant prospect.

I was assigned to the middle of the "stack" because it was my responsibility to protect the team from the half-dozen angry pit bulls surrounding the rundown shack during the approach. The "stack" is a single file formation that is often used at a breach point, like a front door. DB had suggested I bring a short-barrel pump shotgun loaded with buckshot as "bite repellent," but I told him that I'd rather have a 30-round burst of 10mm than just five shells of buckshot if I had to face an angry attack dog. The brave operator assigned to be in front of me in the stack had a fire extinguisher for spraying the bloodthirsty dogs to try to dissuade them from eating us. I was his failsafe. DB surprised me when he gave me his MP5, but I was thrilled to carry it.

"Thirty seconds!"

I felt a combination of mild fear and excitement as the structure came into view from my position behind the driver's seat. The house glowed in the early twilight. In the yard, the pit bulls were already working themselves into a rage. Adrenaline surged in my veins. I felt so alive.

I pulled the side door open on the old panel van, and it locked open with a shudder and a clank. This old van was our "operator delivery vehicle." It looked like an old plumber's van with its flaking paint and black spray-painted ladders strapped to the top. Unfortunately, this and an old white Suburban with custom-extended running boards were the only vehicles our team had. There was no sense in keeping the side door closed on the van now. The door wouldn't stop a bullet, and I'd rather see the fight coming. Our teammates standing on the running boards of our other vehicle, the Suburban, were more exposed than we were.

The brakes on the old van squealed. *Man, I hope they didn't hear that in the house.* Surprise was a key element in our tactics. This was the first round of arrests in Operation Heat Wave. We were rounding up the most notorious drug dealers in the western part of the state, and they wouldn't hesitate to shoot it out with us.

I suddenly felt overly aware that my armor was built to only carry one ballistic plate in the front, and that slot was vacant. We wore the same body armor that the U.S. Army's Combat Application Group (a.k.a. "Delta Force") wore, and surprisingly, it was designed without the ability to insert an armor plate in the back. Hard plates were $800 each, and management wasn't paying for that. My soft Kevlar armor, even without the plate, would at least stop all pistol and shotgun rounds. My surplus Vietnam-era helmet wasn't rated to stop much more than a glancing pistol round. Its leather head strap dug into my forehead and gave me a raging headache within about thirty minutes. I already couldn't wait to take it off. *Let's hope these guys don't have rifles.*

Our Senior Team Leader (STL), Big Toe, was among the operators riding on the skids of the Suburban. Big Toe was a broad-shouldered Army veteran and graduate of Ranger school. He had a legit Tom Selleck mustache and a deep, infectious laugh. His call sign comes from a sarcastic line in the movie *Stripes*: "An army without leaders is like a foot without a big toe." To this day, I have never met someone with a more steady, amicable personality. He was always mellow, regardless of how crappy or how fantastic circumstances were. He exuded confidence and kindness.

"Execute, execute, execute," came his calm order over the molded earpiece in my left ear. I had a foam ear plug in my right ear, but decided to pull it out. I wanted all the situational awareness I could get. The availability of active hearing protection headsets was several years away. Even if we had had them, they wouldn't have fit under our old helmets. Everyone started moving rapidly out of the van to meet up with the operators from our Suburban. I took a deep, slow breath and jumped out of the van.

ANGRY FUR MISSILE

I walked swiftly, with my carbine at low-ready, to my place in the middle of the stack and followed the operator holding the fire extinguisher toward the front door. Since we didn't have armored vehicles, we were vulnerable, so we relied on speed, surprise, and violence of action to protect ourselves. I could make out at least three pit bulls pulling with all their might against their chains and barking frantically. *Thank God they're chained up. This would be an awful mess if they were loose.* We were twenty yards from the door when I saw a behemoth pit bull closing on us from the left of the stack. I hadn't seen this one until now.

I called out, "Dog, left!"

The operator in front of me was going to be blindsided and his fire extinguisher wasn't going to help us. I swung my MP5 toward the dog and quickly ripped through the selector settings on the submachine gun just as I had been trained: *first click - semi-auto, second click - two-round-burst, third click - full-auto.*

Oh, hell…. The dog was only a few yards away as I put pressure on the trigger while still moving with the stack toward the drug house. I could see its teeth. *I'm not going to get a shot off. This is gonna suck.* A voice behind me yelled, "Shoot, damn it!" as the dog leapt into the air.

Clack! The dog's chain ran out of slack just a couple feet away from my face. It yelped and recoiled backwards as if the chain were spring-loaded. I was surprised that it didn't break his neck. He gathered himself and lunged again, but we were still out of range. By the time I moved my eyes back toward the front of the stack, DB, Big Toe, Mag, and Lil Toe were at the door with a ram and a shield.

DB banged on the door and yelled, "FBI! Open the door! We have a warrant!"

I could hear yelling inside the building. The shield-man in front of the stack was protecting the breacher, but he wasn't providing me any cover at all. I felt naked – like a bullet could come zipping through the door or the window and slide through my neck or femoral artery. My heart pounded, but the sensation of fear was long gone. I was energized. It seemed like we were standing on the porch for an eternity. "We're compromised," DB said to Big Toe. Lights flicked on in the house as the yelling continued.

"Breach it," Big Toe said calmly as he gestured to Lil Toe.

Lil Toe had been the most junior operator on the team until I was selected. He was a couple years older than me, and he was pretty reserved with me. I wasn't sure if it was just "new guy" treatment or if he simply didn't like me. He was short and muscular with a full head of hair and always wore a trimmed goatee. He was notorious for giving people "the look" when they annoyed him. He had been adopted as an infant by a loving couple and raised in Kansas where he served as a street cop and SWAT officer. His reputation in the FBI was as a hard-working Agent who kept busy lining up and taking down online child predators. He actually didn't have a call sign yet at this time, but he would earn it eventually after becoming Assistant Team Leader under our Senior Team Leader (STL), Big Toe. When Big Toe would step down from STL, he would stay on the team as a medic and keep the moniker, Big Toe. Likewise, when Lil Toe would step up to being STL, he would still be Lil Toe. It didn't hurt that Big Toe happened to be tall and Lil Toe happened to be short.

Lil Toe took one big swing with the ram. *Crack!* The door frame split at the lock and the door snapped wide open. I leaned my head out to see around the other operators in the stack. I could see deep into the first room in the house, as well as past it, into the far living area. Beady eyes flickered from sunken eye sockets as lights from our carbines lit them up. *Look for guns. Look for hands. Breathe.*

DB already had a flashbang out, and he tossed it through the door the second it opened. The sound from the flashbang seemed less loud than in rehearsals, but I felt the concussion from it in my sinuses. My right ear squealed from the explosion. *I should have left that earplug in.* I was so locked onto the eyes in the far room that I made the mistake of not looking away from the grenade as the magnesium core flashed like a lightning bolt. As I entered the threshold, I tried to ignore the green blob that was temporarily burned into my retina from the flashbang. *I know better. I won't make that mistake again.* The stack moved forward smoothly and began clearing the front room. We were safer inside than we were standing on the porch, and I was happy to get inside where I could at least see who I was fighting.

The first operator entered and pivoted right to clear behind the door. The second operator moved forward, like a human shield, and blocked the entrance into the next room. I cleared the left corner and then moved toward a closed door that looked like it might lead to the garage.

"Closed door left. Need three," I called out to let the team know that it would probably take three of us to clear this area. They knew I would hold this door while I waited for two additional operators to clear the unknown space behind it. I checked the hinges – it was a pull-to-open door. I didn't want it to surprise me by opening into my face, so I wedged the steel toe of my left boot into the gap under the door. I kept my head forward, concentrated on my peripheral vision and resisted the urge to turn my head.

"*Don't touch that gun! FBI. Hands up!*" I could hear DB and Big Toe in the next room giving commands.

"Closed door left. Need three," I repeated, but there was no reply. I had that thought again: *Armor plates would be great to have right now.*

"Cuffing." I heard DB say from the adjacent room.

"Closed door left…" I started to repeat.

"Covering," Big Toe answered him, cutting me off.

I was relieved to hear that the subjects were cuffed. My goggles were starting to steam up. *Breathe slowly.* It wasn't that hot, but I felt sweat running down my head under my helmet.

"Hold watcha got," DB said from somewhere behind me. I was happy to hear his voice in the same room.

Footsteps shuffled up behind me, then I felt a firm squeeze on my left shoulder as DB said, "Got three." That meant that not only was he behind me and ready, but he had a third operator behind him. I stepped to the left and DB opened the door so I could keep both hands on my weapon. The room was dark, but I could see from the light on my carbine that we were in a garage. The garage opened up to the right, so I pivoted right to cover the largest part of the "unknown." Per protocol, DB would be pivoting left, and the third operator would be covering straight ahead. I could see their weapon lights scan their sectors in my peripheral vision.

"Clear right," I called out.

"Clear left," came from the left.

"All clear," came from the center.

We reversed our direction and stacked back into single file to re-enter the main structure. DB put his fist into the doorway with his thumb up like a hitchhiker. "Coming out!" he called out.

The room smelled like garbage and burning hair. There was a black mark on the nappy shag carpet where the flash-bang had detonated.

"Building clear," Big Toe called.

Operators escorted the subjects in handcuffs out of the structure and seated them in the back of police squad cars for transportation to the local police station. After being processed locally, they would be moved to the U.S. Marshals Office to be finger-printed and photographed before facing FBI Drug Trafficking charges in federal court. One of my teammates was the Case Agent on this case, and he worked closely with the state police and the local DEA. There was no animosity between the agencies on cases like these. Illicit drugs like methamphetamine were running rampant in our state, and we were happy to work as a team to round up these traffickers.

STOLEN PLANES AND DRUG TANKS

My partner, Mag, and I stuck around to help the Evidence Response Team conduct the search, but afterwards we hurried back to our tiny three-man office in Norman, Oklahoma, known as the FBI Norman Resident Agency (RA). It was tucked away on the second floor of a strip mall in town. It was a space just big enough to fit our computers, copier, shredder, radio equipment, and critical items to conduct investigations.

Mag was a tall Choctaw Indian, a former Greenville police sergeant, a gifted musician, and an Olympic-class runner. He had been ten seconds shy of a four-minute-mile while he was at the Citadel. I knew I was fortunate that Mag was my partner, my assigned training Agent, and also my mentor. He ensured that I had successfully completed a number of tasks during my eighteen-month probation, such as testifying in court, making arrests, conducting a search, and assisting in a wiretap. Besides assisting with this mission, Mag and I had cases of our own to solve back in our four-county territory.

Among my current fifteen cases, one was an Interstate Transportation of Stolen Aircraft (ITSA) case that involved a drug-smuggling operation using spare fuel tanks in small aircraft to hide drugs. No one inspected or thought twice about what was in spare tanks. As long as you had enough gas to make it to the next airport, you could fly on three of the four tanks and stuff one full of drugs. I had been meeting with managers and students at Airman Flight School and recruiting human intelligence assets at the airport to provide information on the ITSA case.

I was thrilled to be on SWAT, making high-risk arrests, but the core mission of the FBI is the Case Agent's job: to find the truth and document it. SWAT guys had a reputation of being the best Case Agents in the division, and my goal was to make good cases first and foremost.

Mission #6
September 2001

EVERYONE REMEMBERS WHERE THEY WERE

I had spent the last four months eagerly solving "cold" fugitive and bank robbery cases and opening new criminal investigations in my territory. The Operation Heat Wave arrests had occurred just four months ago, and I was drinking from the firehose of tactics I was learning from SWAT. I had a few more SWAT missions under my belt now, too. One of them had been a high-risk drug raid and another was a full day of being "jocked up" (suited up in our armor and gear) and on stand-by for the scheduled execution of the Oklahoma City Murrah Building domestic terrorist bomber. Some militia groups had threatened retaliation on the day of his execution. We were on standby to protect citizens and make sure the bomber made it to his lethal injection as scheduled. The Oklahoma City bombing was personal to our team: Mag's cousin had been murdered in that blast, and our team had responded to the carnage, rescued survivors, and collected body parts.

At the moment, I was at my desk, finishing a final draft for a fugitive warrant. I looked at our government-issued wall calendar and I typed the date for my signature line: September 11, 2001. My gaze drifted back to the calendar, then over to Mag's little black-and-white TV that he kept in the common area of our RA. The news was showing an airplane hitting one of the Twin Towers. Mag adjusted the rabbit-ear antennas on the TV.

Then a second aircraft hit the second tower in New York City and smoke billowed out of both of them. *Is that a replay? That's not a replay. That's a second aircraft.*

"Mag, we're under attack." I said, stunned.

Our pagers vibrated in unison. We had orders to come to HQ. We didn't know what we would be doing, but the SWAT team had been activated.

Early reports indicated that one hijacker, Zacarias Moussaoui, had not boarded his ill-fated flight and had recently been placed into custody by our Minnesota FBI colleagues. Moussaoui had been a student at Airman Flight School at the little airport in Norman, Oklahoma, where I was currently investigating stolen aircraft and drug smuggling.

Mag and I hadn't been at FBI Division HQ in Oklahoma City long before we got a page with orders to return to Norman immediately. Some of our HQ Agents were on the phone with the U.S. Attorney's office swearing-out a warrant to search Moussaoui's apartment in the university's foreign student housing complex in Norman. Mag and I drove straight to the apartment and met with our teammate, Moon, for a briefing. Moon was an energetic and cocky Asian-American with a wry sense of humor and a big laugh. He was the orphan son of a first-generation Chinese immigrant, had enlisted in the U.S. Navy at the age of seventeen, was a former street cop, and an outstanding marksman.

We stacked outside of the first-floor door of the apartment complex. Using a key we had obtained from the university, Moon slowly turned the lock.

I squeezed his left shoulder to let him know I was ready. "Got three," I said.

Once the door was unlatched, Mag announced, "FBI! Search warrant!" He kicked it open, and Moon and I followed closely behind him. There was no one in the building, as suspected. There was a computer set up with a flight stick and a flight simulator box showing an image of a jumbo jet flying past a skyscraper. A prayer rug was laid out. The only shirts left in his closet had screen-prints of the subject's face, overlaid to look like it was on the jumbotron in Times Square. It was surreal. He had left these items and gone to Minnesota to attend a new flight school, where he would learn to fly a jumbo jet, but didn't bother to learn how to land it.

Once we finished clearing the apartment for threats, Evidence Response Team members and Special Agents from the CI/CT (Counterintelligence and Counter Terrorism) squad were waiting outside. They needed to collect evidence of the plot that was responsible for the still-burning airliners in the Twin Towers, in Pennsylvania, and at the Pentagon. The uniforms, training, and tactics of FBI SWAT were about to be substantially changed.

SLEDGE HAMMER

FBI SWAT was formed in 1973 as a domestic national counterterrorism unit and a law-enforcement special operations entity after the Munich Olympics massacre in 1972. The US Army's elite counterterrorist unit, Delta Force, would be formed five years later in 1978 and the FBI's Hostage Rescue Team would stand up in 1983. Even though FBI SWAT was already technically a counterterrorism team, our training and operations prior to 9/11 had been primarily conducting high-risk warrants for criminal cases. Oversight and funding for the fifty-six teams at that time was the responsibility of fifty-six individual Special Agents in Charge (SACs). As a result, FBI SWAT teams were similar and affiliated, but independent. But in 2001, all U.S. tactical counterterrorism units began to restructure to face the rise in terrorism, although our missions were rarely spoken about publicly. I didn't know it at the time, but I was about to witness a new era in tactics and equipment in the FBI SWAT program and take an amazing ride over the next twenty-one years as a SWAT operator.

Instead of fighting with supervisors to get two training days per month like we did in the early days, eventually we would be directed to conduct training a minimum of four days a month, in addition to prepping and executing operations. Once I became the Sniper Team Leader, I saw to it that snipers got a fifth training day. In days yet to come, it wouldn't be uncommon for senior SWAT operators to spend two weeks of each month on training and ops. This was the beginning of the change from being fifty-six loosely-affiliated field SWAT teams into one unified national SWAT team with 1,000 members, stationed in fifty-six different locations. We'd have one common uniform, one common patch, a common motto, common frequencies and crypto, and common tactics.

Instead of being fifty-six independent hammers for our field offices, we would become one immense sledge hammer to protect US citizens from domestic and international terrorism, as well as from criminals. SWAT Basic training would be combined with WMD (Weapons of Mass Destruction) training, so that even the newest operator would be able to operate in a poison gas or radioactive environment. I would see the implementation of armored vehicles, advanced tactics, improved weapons, active hearing protection, air support, tactical robots, upgraded night vision, and mini-drones. Hell, I even got two sets of hard plates.

This was just the beginning.

Chapter 2
Carney Hostage Rescue
Mission #30
February 2007

My huge SUV down-shifted violently as I accelerated up the hill and the speedometer needle hovered at 100 miles per hour. I pushed it as fast as I dared. Hostage situations like these got my mind racing.

There was a state trooper and a young FBI Agent in blue body armor standing at a roadblock. I eased on the brakes, turned off my siren, and rolled the window down.

"Hey, fellas," I greeted. "I'm headed to the SWAT rally point." I flashed my badge and credentials out of courtesy. I was already in uniform.

"Take a right at the bottom of this hill. You can park there. Be safe," the trooper said.

"Thanks, brother." I put the truck back in drive and gave them a nod. "You, too."

I proceeded through the woods, emerging into a grassy clearing, and found a cluster of parked vehicles. Several of my teammates were already here. The mobile Command Post (CP) RV was already set up and FBI supervisors walked in nervous circles while talking on their cell phones.

Unfortunately, Big Toe, our Senior Team Leader (STL), had been stolen from us by one of the Assistant Special Agents in Charge (ASACs). The FBI managers had called him into the large RV and absorbed him into the CP. When they got their hands on him, they rarely let him go. Managers wanted Big Toe at their fingertips, which meant we would hear him on the radio, but we wouldn't see his face until the crisis was resolved. What we needed was a mobile CP that included a functioning Tactical Operations Center (TOC) staffed with Special Agents and Professional Support employees who would manage communications, medical, and air assets and brief managers for SWAT – not a kidnapped STL. In the absence of a TOC, Big Toe would stay in the CP and our Assistant Team Leader, Lil Toe, would be leading us for whatever happened next.

I pulled my body armor out of my truck, tore the Velcro open on the right side, and stuck my head and left arm into it. I took a big breath and cinched the armor closed. It needed to be tight enough that it wouldn't jostle when I was running, but not so tight that it would keep me from taking deep breaths. I grabbed my rifle and sling-shotted a round into the chamber.

Before I put on my headset, I checked that it was plugged into my radios as well as the communications box on my chest. The comms box was the size of a deck of cards and had two press-to-talk buttons, one for each radio on my back. I set my helmet lightly over my headset. Our new helmet was a huge improvement over the old one. It actually fit my head, had padding, and had a mount for my Night Observation Devices (NODs), but it wasn't designed to include a cut-out over the ears for my headset. As a result, the weight of the helmet rested on the earcups of my headset and pushed the temples of my eye protection into my skull. I expected a blinding headache in about an hour. There was no reason to start that timer until I had to. With my helmet on, I let the chin strap dangle free for now and walked briskly toward the Command Post trailer.

More cars were pulling in, and more operators were arriving and getting jocked up. My teammates were driving in from offices around the state. This wasn't a scheduled op, so my guys were racing toward this town to rally with us. Some of them had three-hour drives and some of them had thirty-minute drives. We wouldn't have time to wait for all of them. It was time to rally.

Big Toe emerged from the Command Post. He already looked like he had a thousand things juggling in his mind. He pointed his finger in the air and made a circle with it. Everyone walked over to him and quietly formed a circle with Big Toe at the 12 o'clock.

Big Toe's baritone voice broke the silence. "Monday morning at 5 a.m., the subject, Tony, drove to Kansas, tied up his ex-wife and one of her daughters and kidnapped his former stepdaughter, Annetta. Tony drove Annetta back here to his acreage against her will. He left the two other women tied up back in Kansas. The hostage has been holed up with him in a cabin just over that hill." He gestured with his right arm in the direction of Tony's little house. "Tony is mid-forties, white male, and is said to use a little meth from time to time. The subject doesn't have a cellular phone and he's not answering his landline.

"We don't have the whole team here yet. More guys are on the way, but we're going to prep with what we've got. Ideally, we would set a perimeter, deploy four sniper teams, and assemble an assault team for a hasty hostage rescue. But we don't have those kinds of numbers. Wylie, JimBo, and Mackey, you're going out as three one-man sniper teams. I hate to do that, but we need coverage from three angles and we can't spare any more operators for spotters. Lil Toe, I need you to take everyone else and come up with a hostage rescue plan. I already talked to our sister teams in Dallas and Kansas City and they are on standby. If this goes long, they will be up here as soon as they can to help out. The troopers will have a helicopter here soon. If you need me, I'll be locked in this RV with the ASACs. Be safe."

SIERRAS DEPLOYED

Wiley, Mackey, and JimBo got their bolt-action precision rifles and hiked to their designated hide locations. Mackey was a lean, tall Maryland boy with an accent like no other. People didn't know if his accent was Australian or Cajun, but it was always charming. He smiled and joked a lot, but he was also one of the smartest people I knew. Before becoming a Special Agent, he was a GS-14 supervisor at FBI HQ working in the Computer Analysis Response Unit. Prior to that, he was a network administrator for a large company contracting for NASA.

I had a quick flashback to a day when we were eating pizza after firearms training and I made a joke that no one would know the cube-root of twenty-seven off the top of his head. With a mouthful of pizza, Mackey piped up from the opposite side of the table, "Three." Everyone just stared at him. He was right. After a few jeers about his nerdiness from other teammates, he shrugged his shoulders and took another bite. He was the rare combination of intellectually bright, but still funny, athletic, and he had rock-solid common sense. He earned his call sign after telling Lil Toe that "Mackey" was his father's childhood nickname for him. Once Lil Toe started calling him that, it stuck.

While the snipers were stalking into place, Lil Toe gave those of us on the assault team some intelligence on the layout of the hostage taker's lair.

He began, "There is an entrance on both white side and red side…"

Suddenly, Wiley's voice interrupted us via a dozen headsets simultaneously: "OC-2, Sierra Two." We all stopped and listened.

"I'm in the final OP (observation position) on the north side of the house," Wiley said. "I have eyes on the primary entrance to the residence on the west side. The subject's truck is parked out front."

Lil Toe made eye contact with me as his hand moved toward the transmit button on his body armor. "OC-2 copies. We will be moving out soon. Stand by."

In a perfect world, we would have one team preparing for a hasty hostage rescue and another team preparing for a *deliberate* hostage rescue. In the absence of manpower, we only had the time and operators to prepare the hasty plan.

"J-Money," Lil Toe said to me, "You'll be on point, I'll breach. Let's walk through it." This was one of many assaults I'd done with Lil Toe. At this point most of it was muscle memory.

I had earned my call sign by accident, which is usually the case. I often signed my e-mails to the team with obnoxious names and "J-Money" had unfortunately stuck. I'm lucky my previous signature lines, "Lil Stank" or "MC Def Freshness" hadn't stuck first. We used call signs regularly – and they rarely sounded complimentary, even if they were terms of endearment. If you were holding out for "Razor" or "Hammer," it wasn't happening. If anything, a call sign was a way of checking an operator's humility and willingness to submit to the pack. Even so, when someone called you by your call sign, it was a sign of brotherhood and respect.

As we shuffled through our rehearsal, a popping noise echoed in the distance.

"Did you hear that?" I asked no one in particular. "Was that a gunshot?" I felt a surge of adrenaline.

Lil Toe nodded. His forehead was wrinkled as he squinted into the distance.

"Assault team, emergency hostage rescue is a *go*. Execute, execute, execute." Big Toe said clearly and calmly in our headsets.

HOSTAGE RESCUE

I rode outside of the Suburban on the oversized front-right steel running board. My gloved left hand was on the tubular rail on the top of the truck. My rifle was slung at my side and my .45-caliber pistol was in my right hand so I could shoot from the truck if needed. The truck slowed as we approached the main entrance to the small, rotting cabin. I holstered my pistol, pulled the stock of my rifle into my right shoulder, and jumped off the running board just before the truck stopped.

We had practiced Hostage Rescue (HR) in schools, on buses, stadiums, jumbo jets, and commuter trains; so an HR in a stand-alone residence was one of the simplest venues. The tempo was going to be faster than usual and tactics would be riskier. The HR mindset was less *safety* and more *sacrifice*. With hostage rescues, there was no shield carrier, so there was a chance you were going to catch a bullet in the face. But the tactics had to be aggressive – someone's life was on the line, and it was our responsibility to save her.

We walked briskly, using long, smooth strides, to move to the main entrance of the old wooden cabin. The door had patches of exposed wood where the paint had long since worn away. Lil Toe was beside me with the ram. This old door would be light work for him. The hinges were on the left, so I moved to the right to give Lil Toe room to work. I set my focus past the door, as if I were already looking into the room. My thumb rested on the safety of my rifle.

With a great *CRACK!*, Lil Toe smashed the wooden door free from its frame and it snapped open. My feet were moving. *Living room. Corner-fed left. Sofa in the middle. Kitchen on the far left along with a long hallway. What's the highest threat?*

"Hallway left!" I called out. "Fight left, holding right." I cleared my sector, then I stood in the kitchen entrance to shield the rest of the team with my body. They moved left behind me and worked the hallway while I blocked for them. No person, and no bullets, were coming out of the kitchen without coming through me. My heart thumped, and I took a long breath. It smelled like dirty clothes and cat urine. Long brown cobwebs slung low from the ceiling like cave stalactites. It felt like my headset was crushing my skull. Sweat dripped off my eyebrows and onto the inside of my eye protection. I was hot, my mind was racing…and I felt electrified. I belonged here. In this moment, in this beautiful organized chaos, in the danger. With my team. I belonged.

A squeeze on my left arm released me to move. "With you, Money." Without looking back to see who was with me, I cleared the left corner of the kitchen while a teammate cleared right.

"Kitchen clear!" I called out. Every second we took was a second the subject had to contemplate killing the hostage.

Lil Toe spoke loudly, "Fighting left!" I moved out of the kitchen and linked up with Lil Toe and the team in the hallway. I nearly tripped over a pile of trash and dirty clothes as I closed the gap.

"With you," I said to Lil Toe as I bumped the toe of my boot into his heel to let him know I was immediately behind him.

While I had been clearing the kitchen, they had cleared the first of two bedrooms. All but the last room was cleared. He had to be holding the hostage here.

Lil Toe didn't have to give an order to clear the last room. We were all tracking. The door to the last bedroom was open. I tried to envision where I would be in the next few seconds. *Hinges on the left, so the door opens left. Bedrooms doors are adjacent, so it's a corner-fed heavy-left room. First man will run rabbit, second man will criss-cross to the far corner. I'm three, so I'll clear the center.* In the controlled mayhem of the moment, Frosty announced, "Bang out!"

Frosty was a muscular guy with a bald head and an intimidating look. He was a former Federal Probation Officer and one of our team's breachers. His early call sign had been "Fresh." It changed for little other reason than Wylie once called him "Frosty" and we all joined in. He pulled the pin on one of his flashbangs and tossed it over our heads. The bang hit the open door frame and ricocheted into the room the same millisecond that the team flooded in. I cleared the center and then scanned the entire room – but it was empty, too.

"Clear!" they called out.

"Secondary search," Lil Toe ordered. Everyone retraced his steps and looked for hiding places he could have missed before. I slung my rifle, pulled out my .45 and held it close to my chest. If someone lunged out of a closet, attic, or crawl space at me, I could "make distance" with my empty left and still shoot with the pistol in my right hand.

As I walked back toward the kitchen, I saw a dusty gun cabinet in the hallway. It was open and there was a clean spot where one rifle had been recently removed. I clicked my radio mic button.

"Command Post, OC-3. Dry hole. No hostage located. It looks like the subject is on the move and took a rifle with him." I knew the snipers heard my transmission, and their role was about to change from Overwatch to Reconnaissance. They wouldn't be protecting us for much longer; they were about to start hunting. With no spotters deployed, they had to be looking over their own shoulders with every step, though.

After secondary searches were done, Lil Toe opened the back door of the cabin and we stepped out into the woods. It felt eerily more dangerous outside than it did inside. The hostage-taker knew the layout of his own land. I couldn't help feeling like his crosshairs were on me at that moment. Despite not seeing him, I felt some comfort knowing that Wylie had us covered from an elevated position – for now. Wiley was a former Boulder police officer, father of two boys, and a hell of a guy. You'd find him in denim jeans and cowboy boots when he was off duty. His nickname had followed him from his last job. It was a rarity to keep an old nickname, but it fit him so well that it stuck.

GHILLIED UP AND ALONE

"OC-2, Sierra Two. Contact! White male in a white t-shirt," Wylie called on the radio. My heart rate picked up. I pulled my rifle tight into my shoulder and scanned the wooded hills all around us.

Lil Toe keyed up his mic, "Sierra Two, state the location." There was a painfully long pause. *Where is this guy? We're out in the open!*

"Guys, let's get on the truck and go," I said.

"It's a llama," Wiley said bashfully.

"Say again?!" Lil Toe asked.

"It's just a white llama," Wiley said. "No threat located."

Lil Toe and I made eye contact. He shrugged. This would be a story for later, for sure.

We quickly climbed back onto the Suburban and drove to the staging area beside the CP trailer, where we huddled up to decide our next course of action. The hostage was supposed to have been in the cabin. If she and the subject were gone – and he had taken a rifle – we needed another course of action right away.

A Highway Patrol chopper was thumping overhead now. Mackey's voice piped into our earpieces urgently: "Comman' Post, Sierra Three. I need dat highway patrol choppa vectored to my location right now! I'm in a clearin'. I got da hostage in sight." It was impossible to disregard Mackey's distinctive accent.

From Mackey's perspective, he could see a scared female who matched the description of the hostage, standing on the edge of a wood line across the clearing.

He still had his transmit button pushed. "Come to me. C'mon now," Mackey urged.

I clicked my transmit button. "Sierra Three, this is OC-3. Say again?" I asked him to repeat.

"I see da hostage. Come get me!" he said. His voice dropped in tone and increased in intensity.

"What's your twenty?" I asked, meaning his location.

His radio keyed up, but only static came out of my headset while he held his transmit button down. "Uh." Mackey's mind was working. "We in a clearin' unduh this powah line dat cuts 'cross the property."

"I need you to narrow that down, brother." I said. "Give me another landmark so I can re-section your location."

"Come get me." Mackey said again. I could vaguely hear a man's voice in the background in his transmission. "He still out here. I gotta go." The CP relayed to the Highway Patrol chopper to be looking for one of our snipers in the area. With Mackey in his camouflage ghillie suit, it was like looking for a man-sized bush in an acreage of identical bushes.

Once Mackey was confident that the helo had a visual of him, he keyed his radio. "Comman' Post, Sierra Three. That bird's overhead. Imma drop in the tall grass. Have 'em hover over me." Mackey hoped the enemy rifleman in the woods wouldn't see him blend into the field and that the aircraft's rotor wash would distract him.

As Mackey vanished into the tall grass, he heard the subject's voice call from inside the tree line.

"I know there are snipers everywhere," the subject said. "I'll let her go if you don't shoot." The leaves were crunching under the hostage-taker's feet. The subject probably saw a "moving bush" and convinced himself that every bush was a sniper. Either way, Mackey seized the opportunity.

"Send her to me." Mackey said.

Lil Toe had walked over to the CP trailer to talk though the situation with Big Toe. Time was wasting. I let my eyes trace the power lines overhead. I could hear the helicopter in the distance.

"Lil Toe, I can find him!" I insisted. "I'll get us to Mackey."

"Do it," he answered.

I jumped into the team's Suburban and turned the key. The engine fired to life.

"Mount up!" I called out. Other operators piled in or climbed onto the running boards. We splashed through mud as I drove toward the power lines through an open field and onto an old dirt road. The tires slipped in the mud and sprayed out from the wheel wells. I eased off the gas and shifted into four-wheel drive. Once I felt the front tires start biting, I gave it more throttle. With six kitted-up men on the running boards and two inside, the truck's excess weight wasn't helping our effort to move through the muddy field. Once we were under the powerlines, I turned right and began to scan the horizon. Mackey's camouflage ghillie suit would make him nearly invisible. I was worried I'd miss him or, worse, run over him.

Luckily the hostage wasn't camouflaged. Out of the left treeline, a woman was running toward the powerlines. She ran up to a "bush" and hid behind it. I drove straight to them and said, "Guys, get ready. The second I stop, set security 360. I'll get Mackey and the hostage into the truck." I scanned the treeline to the left. I couldn't see the subject at all.

As I hit the brakes, my teammates jumped off the skids and formed a circle around the truck, kneeling with rifles pointed in 360 degrees. I slammed the truck into park and ran to Mackey with my rifle at low-ready.

"Let's go," I urged. It felt like a bullet would hit me in the head at any second. *If he kills me immediately, I won't feel any pain. If I feel any pain, that means I'm still alive. Just keep moving.*

I ran up to the girl that Mackey was shielding with his body. He was scanning the woodline for threats through his scope.

"Annetta?" I asked her. She looked up at me with big eyes. Her pupils were dilated from fear. She was completely overwhelmed. She stared, paralyzed with anxiety, at the dozen armored men with rifles scanning the horizon for her attacker. I put my hand out for her to take. "Come on. I got you." I said calmly. She hesitantly grabbed my hand, and I led her into the back of the Suburban and shut the door.

"Recover!" I yelled to let the guys know it was time to go. I jumped into the driver's seat and wedged my rifle under my left arm.

"Last man!" An operator standing on the skids thumped on the roof of the Suburban twice. I shifted into drive, stomped the gas and mud flew behind us as the tires broke loose. I spun the truck 180 degrees and headed back to the staging area. Once we were a couple hundred yards away – far enough away to be hard to shoot at – I slowed down. It felt like I had been holding my breath the entire time.

When we got back to the CP, we walked Annetta to a waiting ambulance where EMTs looked her over to make sure she was unharmed. Case Agents stood patiently a few feet away, waiting their turn to ask her for the details of her abduction. Meanwhile, the armed hostage-taker was stalking through ground that was familiar to him and foreign to us.

MAN HUNT

We quickly switched gears. We set wide isolation to make sure the subject could not escape the area. State and local police were still arriving, and we formed multiple perimeter and man-tracking teams. The highway patrol tactical team had assembled, and the sheriff's office brought in two bloodhounds to track the subject.

We planned out patrol sectors using our team and the troopers. We had to move fast and keep the subject from escaping. But that meant we were also walking back into the mouth of danger. We were on his property. He knew every good hiding spot, each hollow tree, each location he had used to stalk deer that he could now use to ambush us.

We assembled into a "Y-formation" that put two riflemen out front, and let an operator stay in the middle of the formation to focus on mantracking. I took up the position on the left tip of the "Y." As we moved slowly through the woods, the tracker looked for overturned blades of grass, broken twigs, footprints or any other "sign" or "spore" from the subject. He made hand gestures to us in the rifleman position to change our direction without speaking. We weren't moving silently, but we were certainly using noise discipline as much as we could. This was a technique that would allow us to fight by moving quickly to an "L-formation," but it was anything but safe. In a gunfight, "action" almost always beats "reaction," and we were in a position where we were relying on "reaction." The chances of us taking a casualty were high.

The property was thick with cedars, blackjack trees, and deep ravines. Every time I crawled into a ravine, there was a slight relief because I was low enough that I couldn't be seen from above. However, each time I crawled out of a ravine, the hair stood up on the back of my neck. It felt like his crosshairs were on me the entire time.

This is what it feels like to be hunted.

I imagined what I would do if I was in Tony's shoes and I had murderous intent. It wouldn't be that hard for him to kill at least a couple of us before we could neutralize him. Now that we had him penned in, someone had to go into the woods and find him. There weren't a lot of other choices. The perimeter team couldn't hold forever.

As we were pushing through the woods on the western sector, the highway patrol tactical team was in the eastern sector with the sheriff's bloodhounds. Suddenly a shot rang out east of us and echoed through the ravines. We pivoted and moved quickly but cautiously toward the sound of the gunshot. Moving *toward* the sound of gunfire seems like a crazy plan, but we had to catch this hostage-taker. I hoped the gunshot wasn't the result of a trooper getting ambushed by the subject. Unfortunately, the highway patrol radios were incompatible with ours. We were calling our CP to get them to relay to the troopers that we were coming to them. It was like a terrible game of "telephone" between two tactical teams.

As we arrived in the area, troopers were standing over a body. The subject had shot himself in the head. He was gurgling and gasping. He was dead, but his body didn't know it yet. The troopers called for a medevac and the subject was loaded onto a helicopter and rushed to a hospital, but his brain was already catastrophically injured. We hiked in silence back to the CP.

The hostage was bundled in a gray wool blanket and sitting comfortably on a small chair inside the mobile command post. She was leaning forward taking small bites from a piece of lukewarm pizza. The blanket and the pizza were both courtesy of the local Red Cross. The small-town Red Cross volunteers were great. Ironically, if we did an op in a big town, the Red Cross was never available. In a small town, you could count on them. You could feel the hometown care from these volunteers as they walked around offering food and water to all of us.

For Annetta, this nightmare was over. In the last 24 hours she witnessed her mother get tied up, was driven across two states, holed up in a shack, dragged out into the woods at gunpoint, and rescued by men dressed up as bushes while a helicopter hovered overhead. She was still mentally in shock, but she was safe now.

LINE UP

Transitioning back to a regular work day after an op was always a little surrealistic. I had to drive down to the FBI Headquarters in Oklahoma City the next morning to sign some reports and check in some evidence. People were chatting over cubicle walls and sipping on coffee. Supervisors sat in their corner offices reviewing files at their computers. It was all strangely routine. *I was in a hostage rescue operation yesterday and people are asking me about college basketball. It's so odd.* The folks at HQ were always friendly to me. I usually added an extra hour to my schedule when I had to go to HQ because I knew I'd spend that hour standing in the hall talking to colleagues that I only saw face-to-face every few months.

I passed one of my favorite analysts in the hallway carrying a full cup of coffee in his preferred mug. He smiled kindly and asked, "Hey, stranger! What have you been up to?" I knew I couldn't answer, "Well, I cleared a house searching for an armed hostage taker yesterday. Oh, and

I watched him die. How about you?" I just smiled back and said, "Not much, brother! How have you been!?" After a minute of exchanging pleasantries, we each continued on our way.

Just a few minutes later, I passed Wylie in the hall and there was an instant connection. We gave each other a wry smile. We both had the same thought: *We were in the shit yesterday!* Wylie gave me a tight hug. Some colleagues who passed us in the hall rolled their eyes at us. I knew that they thought it was odd that we were hugging for no reason in the middle of the hallway. I understood that SWAT seemed like a closed private club to some, but it wasn't intentional. It's just that we knew what each other had been through, and we had a kinship that surpassed being just colleagues.

I followed Wylie up the stairs and to the end of the north wing, into the Counterterrorism Squad cubicle area. We asked about each other's families and told funny stories as we walked. Once we arrived at his desk and sat down, we began re-living yesterday's op and dissecting it a bit.

"How was Overwatch yesterday, Wylie?" I asked him. "I sure was happy you were covering us when we came out of that cabin and the subject was already on the loose. That was sketchy."

"I know!" he said, "After you said the subject was in the woods with a rifle, I felt like every tree rustling was him lining up to take a shot at me. I'll carry a pistol with three extra mags on sniper deployments from now on. If I don't have a spotter and I'm on my own, a big bolt action is the wrong fighting tool." I nodded and thought about what that must have felt like for him.

We sat for a moment in a comfortable silence. I noticed there was a manila envelope on his desk that said, *Wylie* in Lil Toe's handwriting.

"What's that?" I asked.

Wylie shrugged his shoulders and opened the envelope. It contained an official FBI six-man subject line-up form. It had photos of five men, each wearing white t-shirts…and one white llama.

Chapter 3
Glass Mountain
Mission #46
May 2011

CALL OUT

My seven-year-old daughter had been put to bed, stories read, prayers said. My wife had gone to bed and I was upstairs, settling into the loveseat and turning on the TV. But when I made my body still, my mind would spin up. My laptop came out and I started setting appointments and reminders for myself: EMT recert, church event, court date, upcoming trips, case work, subpoenas, travel vouchers, scheduled training...

My phone began blaring its frenetic call-out alert. It sounded like there was a police siren in my pocket. *Please stop.* I pressed frantically on all the phone keys at once. *Please don't wake everyone up.* We had our Bureau cell phones set so that when a tactical WARNO (Warning Order) went out, the phone would wail at maximum volume, even if the phone was in silent mode. I unconsciously stood to my feet as I read the WARNO.

The message was from our SWAT Senior Team Leader (STL), Lil Toe. He was a fellow sniper, but didn't deploy as one since he was in the STL position now.

Guys,
Just got call from Spence regarding a Federal A&D fugitive from Jacksonville. Guy detonated a bomb at a place of worship and has

handguns and an AK-47 style rifle. Solid intel he's in Glass Mountain area. Apparently, a wide-open desolate-type area. Don't know if stopped for night or what. Troopers headed into area to see what they can. Appears to be stopped on wide open dead end with good view.
Be ready, we may be getting a call.
Advise if available.
Lil Toe

A&D meant Armed and Dangerous. When the final call out came, I was the closest to the crisis area, yet I still hurried to get into my olive-drab Crye uniform. I kept every conceivable piece of equipment in my old Bureau-issued 4x4 ¾-ton Suburban. Just as a pro golfer has all his equipment in one golf bag, the Suburban had a full EMT medical suite, in addition to a selection of weapons, with ammunition and optics for all of them, plus food, water, batteries, night vision goggles, surveillance equipment, a sleeping bag, radios, a ballistic shield, breaching gear, and much more. Everything had its place. I left a note for my wife and a separate note for my daughter. They both hated to be woken up. Then I headed out.

I waited until I was down the road a bit to turn on my emergency lights, so that I wouldn't wake up my family. Once I got on the highway, I turned on the siren. Then I turned it back off. Listening to the siren for two hours was going to wear me out – and there was no one else on the road. I radioed the FBI dispatchers, known collectively as "220," who were in the Communication Center in Oklahoma City 70 miles away.

"220, OC-2. Show me 10-8 to the crisis site." I pushed the speed up to about 100 mph and settled in. 220 advised that the SWAT team was going to rally at an ad hoc staging area near an isolated grain silo in a small farming community. I punched in the coordinates and pressed on.

Once I was close, I slowed down and scanned the murky night horizon. Then I saw the outline of the grain silo. The crackling of gravel under my tires as I came to a stop at the tower was the loudest thing for miles. There was no one there. I was either lost or I was the first on scene. Regardless, I took the time to pull out my gear and double-check everything: new batteries in the optics, radio charged, press-to-talk in operation, headset batteries installed, new GPS battery. It was amazing how many batteries it took to keep an operator in mission-ready status. All my gear had a faint musk of old sweat and loamy soil. I liked that smell. It's what a good op smelled like.

I did a check of my pre-packed sniper backpack or "ruck": compass, GPS, compact binoculars, multi-tool, night scope, MRE (meal ready to eat), two spare rifle magazines loaded with match ammo, one spare magazine loaded with bonded ammo (marked with blue tape), extra .308 ammunition, signal mirror, medical kit, waterproof notepad and pen, whistle, LED navigation light, batteries, pace counter beads, trauma kit, headset, IR chem stick, face paint, antiglare scope cover, laser range finder, heater packs, bug spray, sunscreen, bandages, rain jacket, IR floodlight, survival kit (matches, space blanket, water purification tablets), boonie hat, ghillie hat, pruning clips, 550 parachute cord, camo netting, wool watch cap, empty camelback, two full water bottles, and a ziplock for my non-waterproof cellphone. This surprisingly all fit in a small ruck that weighed 26 lbs. My rifle and pistol added another 18 pounds, and my armor and helmet 25 more.

Just as I was wrapping up my inspection, a cluster of headlights cut through the moonless night. The distinctive drone of our 15-ton military surplus MRAP (Mine-Resistant Ambush-Protected vehicle) echoed down the road and demanded my attention. FBI SWAT went decades without any armored vehicle protection, so finally having even this old DoD reject from the Gulf War was a treasure. As operators and Electronics Technicians (ETs) poured out of their trucks, there was a crescendo of conversation while they all started their own gear checks.

I was grateful for the head-start getting my own kit set up. I walked over to Lil Toe's truck to come up with an ops plan. After ten years on the team, I was the second- most senior operator. Lil Toe was in charge and I did not envy him. Kicking in doors was straight-forward. Creating ops plans on the fly was hard work. But Lil Toe was a pro and he was up to the task, as always. He showed me the possible location of the subject on his personal iPad.

"Money, I need a sniper team to locate the fugitive." He made a circle with his finger over the iPad. "He should be somewhere in here."

A park ranger walked over to us and introduced himself.

"There are a bunch of tall mesas in the park. If you climb to the top of this one," he gestured with a finger into the inky nothingness of the moonless night, "you'll be able to see for miles. I'll take you over there, if you want."

I saw my sniper teammate, Mackey, doing his gear check a few cars down and walked over to him.

"Mackey, is your sniper kit ready? We're deploying a sniper team to locate the subject. I want you to roll with me."

"Hell, yeah, Money. I'm 'bout done anyway. Where we goin'?"

"Jump in with me. We're following this park ranger."

"Hey, I ain't got Shania. She in the mail to that school I'm goin' to." (Mackey had named his bolt-action precision rifle after Shania Twain.)

"No worries, man. I'll give you my M4 and you can cover our six on infil."

"Sounds good, Money."

Before Mackey and I left, we walked to where the team was assembled. Everyone had instinctively formed into a perfect circle. As Lil Toe spoke, all the side conversations stopped.

"Listen up. The fugitive went off the grid in Florida where he was the primary suspect in a church bombing. A search warrant turned up additional pipe bomb production equipment in his residence. Technical intelligence has put him back on the grid and now it's our job to bring him in. He's probably armed with an AK-47. Double check your gear. Troopers have set perimeter. Once the snipers locate him, we will be prepared to effect his arrest."

RECONNAISSANCE

Mackey and I followed the park ranger to the base of a mesa and loaded up. Just as the park ranger drove away, he said, "Y'all don't get too close to the edges. You'll fall to your death."

I put on my ruck and slung my 24" heavy-barrel bolt action rifle over my shoulder. I checked that my .40-caliber Glock G23 was loaded, chambered, and the weapon light was operational. I took out my compass and GPS and punched in the coordinates that Radio had given me. Radio was an ET that supported the team in addition to his day job of maintaining all FBI radios, repeater towers, base stations, surveillance cameras, alarm systems, encryption and more. Radio was always great to have on a call out. He was a good friend and we had daughters that were the same age. There was always one radio or headset that needed last-minute repair and he was ready to fix or replace it on short notice. He was a critical part of the team. Even though it wasn't his job, he usually brought water for everyone and extra batteries of all kinds.

Once I punched in Radio's coordinates for the subject, we set out. The hike was fairly simple and there was a distinct path at first. As we got to the top, we flipped down our helmet-mounted NODs (Night Observation Devices) and we saw the world in shades of green. We continued our trek without giving off any light emission that might spook our fugitive. No headlamps, no flashlights.

There was a crunch and I heard Mackey grunt as his weight shifted. I instinctively grabbed his arm and pulled him toward me. Rocks tumbled off the side of the mesa next to Mackey's feet. *The park ranger was right about the edges being sketchy.* We had hiked to the edge without realizing it. We could see the ground in the moonless night, but we didn't have any reference to know we had approached the edge. As I scanned the area with my NODs, I looked for any sign of the fugitive below us. I scanned the sky and the Little Dipper jumped out at me. It reminded me of the nights my dad would take me to "rough it" in the Rockies with nothing but what we carried on our backs. The North Star was easy to find in the clear sky. I had been navigating north by compass, but now I flipped the compass closed. The North Star strobed brilliantly, so I used the stars to maintain my course. Better to keep my head up and not walk off the edge of a tall mesa.

As we moved forward, we would stop often to listen and observe. I pulled out my rifle and mounted the PVS-22 Universal Night Sight on the picatinny rail in front of my Leupold Mark-4 scope. I snapped a Steiner infra-red (IR) laser floodlight to the three-o'clock rail. Using these aids, I could scan the entire area below for any signs of the fugitive. Several hours had passed and we were starting to feel the chill and fatigue. We had no idea how long this night would go.

"There," I said, and I pointed my infrared floodlight at an object. The IR floodlight was invisible to the naked eye, but shown like a flashlight through my scope. I knew Mackey could see the laser through his NODs. "Is that a car?"

"Nope, Brotha. That's a pump jack."

"No. Next to the pump jack."

"Bubba, that's a cedah tree."

"Copy. Man, I don't see anything else out here."

I pulled out my range finder and fired the invisible laser at the pump jack. The screen showed 813 yards. I sighed, "Okay. So. There's nothing for 800 yards in all directions. Let's keep moving north until we fall off this thing."

"Like a Wiley Coyote cartoon?"

"Exactly."

Before we moved, we both put on rain gear. There was no chance of rain, but the cold air and wind were making us both shiver. I spontaneously pulled out my camera and took a photo of the two of us – two sniper buddies, cold to the bones, hunting a fugitive bomber from the top of a mesa in the middle of the night. *God, I love this job.*

We hiked 800 meters, the length of the mesa, and stopped at the far north end.

"This is as fah as we go," Mackey quipped.

"OC-1, Sierra One," I whispered into the boom mic on my headset. OC-1 was the radio call sign for the SWAT STL. Sierra One was my call sign as the Sniper Team Leader.

"Go ahead for OC-1" replied the familiar voice of Lil Toe.

"Roger. Sierra One is on station at the north end of the mesa. We are setting up an OP (Observation Point) and scanning for the suspect's vehicle."

I flipped my NODs up, and let my eyes adjust to the dark while I unslung my rifle. Mackey scanned the horizon with his NODs and identified suspicious objects while I zoomed in through my scope to further identify them.

"400 yards out, at one o'clock," Mackey said.

"Hold on." I swung the rifle to the right. "I see it."

"Is that a truck?"

"Negative. It looks like an old oil tank."

After scanning every inch of the land around us, I pushed my boom mic against my lip and pressed the transmit button. "OC-1, this is Sierra One."

"Go."

"OC-1, there's no personnel or vehicles in sight."

"Well, keep looking." Lil Toe sounded annoyed.

"Copy. Will do…. Hey, are we getting any air cover?" I asked.

There was a pause and then I heard the radio key up. "Yeah, Voodoo is on the way. No ETA." Voodoo was the call sign for our local aircraft.

I keyed up the radio again. I couldn't remember Radio's official call sign. *Is it X-Ray 39? 38?* Screw it – he knew his nickname.

"Radio, Sierra One."

"Go ahead, Sierra One."

"Roger, can you send the most recent suspected grid coordinates for the subject?"

"Will do, Sierra One. Stand by."

Within minutes, I got an email from Radio with the most recent coordinates. They were different from the last time. *Maybe they are getting better triangulation now.* I put the new grid coordinates into my hand-held GPS. Once I had a bearing, I used my compass to shoot an azimuth to where the suspect should be. I held up my compass, turned it until the correct heading was at the top, then made note of the boulder it lined up with. Now I could put the compass away and use the boulder for reference.

"Mackey."

"Yeah, man."

"He should be 800 yards that way." I gestured north with my hand.

"Awright."

"So why can't I see him?"

We sat in silence. The sky began to change hue. We watched as the sun cracked over the eastern horizon. *Man, it's tomorrow morning.* My toes were numb from the cold. I wiggled them to try to get blood flowing. It felt like someone was stabbing me in the belly button. I had recently developed an umbilical hernia from squatting way more weight than I should have. It turns out, you can lift with your legs and still rupture your own abdominal wall. I had surgery scheduled for the following morning. Now the pain from my umbilical hernia came rushing in; I had been sitting motionless too long.

Then a glimmer of light reflected off red rock directly to our north.

"Son of a gun," I muttered.

Mackey saw it, too. The sun had revealed that there was another mesa directly in front of us. From our elevation, the top of it lined up perfectly with the dawn horizon, so that in the moonless darkness, it had looked like just a dark patch of Oklahoma prairie.

I grabbed my laser rangefinder and shot the distance.

"700 yards," I said. "It's 700 yards to the south side of that mesa. He should be 800 yards away and that mesa is about 100 yards across."

"He's on the far side of that other mesa."

"Yep."

"Well, let's move," Mackey said with a grin.

In almost perfect timing, I saw a small aircraft approaching from the northeast. I keyed up my radio. "Voodoo One, Sierra One."

No answer. I hoped that he had his radio set to our default Fight-Net frequency.

"Voodoo One, this is Sierra One. How copy?"

The radio crackled to life and I heard the broken transmission of my buddy, Big Al. I could immediately envision his devious grin as he sat in the left seat of the plane. It made me smile and my chapped lips cracked. Al and I had worked violent-crime major-offender (VCMO) cases in adjacent territories of southern Oklahoma when we were new Agents. We had chased bank robbers, drug dealers, and fugitives together. We had a common passion for flying and had started the FBI Aviation program jointly. We ultimately both bought our own four-seater planes and even flew formation together on some occasions. Al was a former Illinois police officer and IRS-CID (Criminal Investigations Division) Agent. He was below-average height, way-above-average intelligence, and bursting with enthusiasm for life. I couldn't help but be in a good mood when I was around him.

"I gotcha 10-2, Sierra One. What's your location?" asked Al. I could hear the grin in his voice.

Before he could finish his sentence, I had my signal mirror out. I had learned how to signal an aircraft during SERE (Survival, Evasion, Resistance, and Escape) training at the US Air Force Academy. At one point in the training, we had to vector in rescue helicopter pilots and signal our location with mirrors, flares, and/or smoke. Once they had our position, the chopper pilots would radio to tell us to knock it off because we were blinding them with our mirrors.

"Is that a strobe, Sierra One? What is that?"

"A signal mirror," I replied. "You have my location at the north edge of the large plateau?"

"Affirmative."

"Roger. Look north to the smaller mesa. The coordinates put the subject on the north side of that mesa. Can you scout for me?"

"No problem, Sierra One. Stand by."

The aircraft rolled smoothly into a northbound turn, and I watched it silently make a lazy circle. If you weren't actively looking for it, you would never notice the little plane cutting holes in the air above.

As I waited, I reminisced about responding as backup to a bank robbery in which Al and his partner, Craig, had chased a subject until the perp was cornered and drew his CZ 45-caliber pistol. The robber had been standing behind a vehicle; Al's partner fired one .40-caliber round over the top of it and hit the robber in the neck. Al had to wrestle with the robber to stop resisting so they could apply first aid to the hole in his neck. When it was all over, the evidence response team took Craig's pistol as evidence. Since managers on scene didn't have a spare pistol available, Big Al gave Craig his own sidearm so that he didn't have to leave the scene with an empty holster. Al wasn't just a pilot; he was a seasoned street Agent who happened to be a great pilot. Warrior first, pilot second.

"Sierra One, Voodoo One"

"Go for Sierra One."

"Tally Ho on a pickup truck on the north side of the mountain."

"Great work, Voodoo One. Thank you." I replied.

Mackey and I scrambled to take down our OP while I called OC-1 and advised him that we would be relocating to the mesa to our north. We hiked back about 1,000 yards to the south side of the first mesa and took the switchbacks down to my Suburban. Spence, a Supervisory Agent and our SWAT coordinator, met us at the truck. Mackey and I lightened our rucks for the steep climb we had ahead. We stowed our night vision, got more water, and inserted Level-IV ballistic plates into our soft body armor. Once we were set, we drove around the first mesa to the south side of the mesa that was masking the subject.

INFIL

Mackey and I stared at the south face of the new mesa.
"Man, that's a steep sumbitch."
"Yeah, it is." I answered. "Let's go around the west side. It'll be an easier hike, and we'll still have good overwatch for the team."
"Nah, man. We need to go straight up to the top."
I grimaced at Mackey. I knew this climb would hurt, but I knew Mackey was right.
I let Lil Toe know that Mackey and I were starting our infiltration to our new observation point. "OC-1, Sierra One. Sierras are starting infil to final OP."
"Copy, Sierra One. Let us know when you're set," Lil Toe answered.
I hoisted my ruck and rifle over my shoulder and started up the face of the mesa. It was more than a 20-story climb and the combined weight of armor, supplies, and weapons was still about 70 pounds. I started out first and Mackey followed me. It was a slow climb because each foothold and handhold was critical. The mesa was a mix of soft soil and a crunchy, flakey white quartz-like stone. I was about five stories up the cliff when my right foot broke loose from its perch. All my weight shifted to my hands, but the soil under my hands was breaking apart. "Mackey, watch out!" I was sure I was going to fall and push him off the cliff with me. I scrambled to find a new foothold, but nothing would hold. This was it. *I'm about to fall off this mesa.* Suddenly, I felt a solid footing, but it wasn't the mesa. It was Mackey's hands. He grabbed my boots and pushed me back up until I could latch back onto the cliff.
"I gotchu, Money."
"Brother, I owe you one!"
"Well, you didn't let me fall off the side of the last mesa. So maybe we even," he said.

Once we reached the summit, I dropped my pack to lower my profile and we low-crawled slowly across the flat top of the mesa. Near the north edge, I saw a dark pick-up truck and a green tent below. We were moving from the reconnaissance (looking for the bad guy) to the surveillance (looking *at* the bad guy) phase of snipercraft. I set up my rifle in a stable firing position and then moved to the side of it. The fugitive didn't know we were here and there was no one that he could hurt, so my primary weapons were a camera, binoculars, and my radio – not my rifle.

"OC-1, Sierra One."

"Go ahead."

"Sierras have eyes on a white male subject in a tent and a vehicle matching the description of the bomber's truck. Photos and diagram to follow."

"Good work, guys. We're working on an arrest plan. Keep me posted."

SURVEILLANCE

"Is he dead, Mackey?" I was laying on my back stretching my neck while Mackey was looking through a magnified optic, or "on glass." The fugitive was lying perfectly still inside his tent. The surveillance phase would be the worst part of the entire mission. Mackey and I traded off watching the subject through binoculars, sketched out a diagram, transmitted images from our phones, and started planning contingencies.

"He ain't dead, Money. He just rolled over to take a piss." Judging by the mostly empty case of beer open by his truck, he may have been sleeping off a hangover.

Well, that at least makes this a little more interesting. But the adrenaline from the climb and locating the fugitive was wearing off and leaving us drowsy. We had deployed last night and were into the second day of the operation. The time was passing slowly and painfully.

Mackey and I collaborated on possible shooting solutions by collecting the wind direction and velocity, as well as the angle down to the subject. The wind was blowing left to right across my face, but the prairie grass by the bomber's tent was moving right to left. I had experienced this before while practicing off-angle shooting in the local university's football stadium. There was no way to know what the wind was doing between me and the fugitive. I decided to hold center windage if I had to shoot. I fired my laser range finder toward the fugitive's tent: 210 yards. That was the hypotenuse distance from me to the target, and while wind does affect the bullet's flight for the full hypotenuse, gravity only acts on the bullet for the horizontal distance over the ground. I used basic Pythagorean theorem to determine my hold-over based on my elevation and 210-yard hypotenuse. The altitude of our mesa was 1,548 feet MSL (mean sea level); subtracting the altitude of the fugitive bomber, I determined the lateral distance to the threat was 197 yards with a 20 degree down angle.

We would be responsible for guiding the arrest team and conducting overwatch while they executed the arrest. More hours passed and the thrill of this mission was fully waning. My fingertips stung. The tips of my gloves had torn away during the climb, and dirt was jammed under all my fingernails. *This sucks. I wish I was on the assault team right now.*

I looked at Mackey, "We're out of water."

"Ahh. I don't wanna hear that," Mackey answered.

Blowing dirt scraped my eyelids and made my tongue stick to the top of my mouth. When it was Mackey's turn "on glass," I would low-crawl back out of sight and lay on my side to urinate. Then I would roll on to my back and rest my neck. Craning my head up to watch a subject was literally a pain in the neck.

I was growing more impatient. I sent a text message to Frosty, now one of our assaulter Assistant Team Leaders.

Hurry the hell up. Drive up on him in the armor. Get him now while he's sleeping.

Frosty sent me a terse reply: *We're working on it.*

This is taking too long.

Luckily, I wouldn't learn until later that on this day I was square in the middle of "Rattlesnake Round-up" territory, and it was prime rattler season. People came from all over the state to compete to see who could catch the largest vipers. Had I known that at the time, I definitely wouldn't have been able to lay flat and relax my neck for fear some venomous reptile would try to smite me for invading his territory.

The plan Mackey and I sent the team suggested that they drive our armored truck far north of the mesa, drive through a barbed-wire fence, and then make an approach straight south toward us and the bomber. They couldn't come from the east or west because there were ravines that looked impassable for any vehicle – even a four-wheel-drive MRAP. I suggested they just drive right up to the tent and call him out on the PA speaker. They would be safe in the truck and there was nowhere he could run where he would be out of the range of my scope, or Voodoo, or the state troopers on perimeter.

OVERWATCH

After several more hours of surveillance, we got the radio call that the assault team was on the move. We were starting the "overwatch" phase, so I made sure my bipod was secure in the flakey white rocks, but not so close to the edge that it might slip off. I leaned my weight into the rifle and the fragile white crystal flakes crunched. *I wonder if that's why they call it "Glass" Mountain.*

I mounted my rifle, loaded a round into the chamber, and made sure the safety was on. The front plate in my body armor made it hard to mount the rifle and keep it oriented down at twenty degrees, so I pushed an MRE and part of my ruck under my chest. My shooting position needed to be durable. A great shooting position isn't great if it's exhausting to hold that position. No one tells you that being a sniper in real life means the boredom of sitting perfectly still for hours, ignoring an itch on your nose, wishing you had more water, needing to take a bathroom break, sweat burning your eyes and dirt up your nose – followed by unexpected, wild bursts of adrenaline.

At this point, shooting was second nature. I got a wave of satisfaction and comfort every time I settled in behind a rifle. I had fired over 200,000 rounds at this point in my career and another 300,000 rounds would follow. The winds were a lot like the day we did live-fire in the university stadium. The angle and the distance were also a lot like the stadium training. It almost felt like just another day at the range. From here in my 'hide', I was invincible.

I used my radio to guide the assault team into position. When they got close to the fugitive, I keyed my mic again. "You're seventy yards from him. Pull straight up." I could barely make out one of our new operators, Vulcher, in the driver's seat, 270 yards from my position. The expression on his face was stoic.

"We can't," Lil Toe answered.

"What do you mean?" I asked.

"It's too steep."

I looked at Mackey and he shrugged. We would later learn that we were experiencing our second optical illusion of the mission. In addition to the mesa confusion, now from our perch, looking down, the grade of the slope up to the fugitive's tent looked shallow, something one could easily drive up. But from his driver's seat, Vulcher was looking at about a 45-degree grade. That's not something you want to risk in a 30,000-pound top-heavy armored vehicle.

The PA speaker of the MRAP crackled to life. "This is the FBI. We have a warrant for your arrest. Come out of the tent with your hands up. We are not leaving." Dangler dismounted and set up from behind the MRAP, and Wylie set up in the turret. The call-out commands to the fugitive were clear, even from my position on top of the mesa. We waited and called out repeatedly for over twenty minutes. We had time to our advantage. There was no reason to march up to the tent and get shot, but there was no way for a vehicle to get to him. He was on the high ground relative to the assault team, but not to me. And he had no idea I was watching him.

"He's moving in the tent," I called on the radio. "He has to hear you."

As a contingency plan, Frosty, South Beach, and Twilight moved on foot to a flanking position to the east of the fugitive. As they closed in, the subject zipped closed the bug flap on his tent and the hairs on my neck stood on end. I watched his tent with my right eye through the scope and kept my left eye open to see the big picture.

"Frosty, stop. Hold up. Don't get any closer." I called over the radio. From my perspective, the fugitive had a perfect shot at my guys if they came through the briars that separated them from his tent. Zipping just the bug flap and not the rain flap had given him the ability to see out, but we lost the ability to see in. It was too dangerous for them to go any closer.

The terrain was too rough for either our little "throw-bots" or the bomb tech's big robot to work, and drones weren't authorized equipment for us yet. A state patrol K9 unit was called in, and he and another trooper approached from the west, but with enough offset to prevent a crossfire with Frosty and the flanking team.

"Headshots are sexy an' all, but if dis guy come out shootin', just put 'im down wit a torso shot." Mackey said. Without taking my eyes off the subject, I answered, "Solid copy, man. I'm on the same page."

The K9 handler released his dog, and it ran for the fugitive's tent. Once the dog was released, I knew it was going to get serious. Now the warm flood of adrenaline was back and the thirty-one hours without sleep didn't feel so bad. I struggled to swallow as my tongue stuck to the top of my mouth. I wished I had brought more water. I wanted to say a prayer, but I couldn't think of anything eloquent. I simply said, out loud, "*God, give me wisdom,*" and I snapped off the rifle's safety.

The dog arrived at the subject's tent but quickly spun around and reversed his direction back to his handler. Because the fugitive was on the north side of the tent, all I could see was his head above the tent-line.

I kept both eyes open so that I had a better perspective of the crisis site. My right eye stayed trained on the head of the subject through my rifle scope, and my left eye gave me a complete view of the dog handler to the west and my teammates to the north.

The dog headed again to the fugitive's tent. The subject suddenly stood and began turning to his left. I could see a rifle come into view from behind the tent. I immediately recognized the rifle as an AK-47. In a millisecond, I thought in succession: *Is that an AK-47 or an SKS? Is that an EOTech attached? How did he attach it? There must be a picatinny rail on the dust cover. Why am I thinking all this right now?*

He was pulling the stock into his shoulder and the muzzle was coming upward and toward my teammates at the MRAP. In a tenth of a second, I thought, *Oh my God. He's going to kill my teammates.* There was no gentle squeeze; there was no holding my breath; there was no letting the shot surprise me.

Time to fire – RIGHT NOW.

I pressed the trigger fully to the rear and let loose the 168-grain jacketed open-tip boat-tail at 2,752 feet per second. I fired the gun just as I had fired a 16-gauge so many times quail-hunting on the Oklahoma prairie with my Uncle Jim. The rifle pushed back into my shoulder but was eerily quiet. This wasn't my first time to fire under stress, so I knew that I was experiencing auditory exclusion. I could hear a chatter of gun fire after my trigger press, but I didn't hear my own shot. My ears wouldn't even ring later, despite my lack of hearing protection on my right ear.

It took the projectile a fraction of a second to cover the 210 yards and land right in the middle of my crosshairs which were resting on the tent-line below the subject's head and where I knew his torso would be. He dropped to the ground while letting loose a full-auto burst from his assault rifle. His bullets made a ripple of blooming red Oklahoma dirt that chased towards my teammates and then stopped abruptly as he collapsed. Those bullets could have torn through our soft armor like so much paper mâché. I cycled the bolt violently while I kept my eyes on the threat. A voice in my head said, *You should cycle the bolt*, and in the same second another voice said, *I already have.*

Mackey looked at me, "Good hit."

I resisted the urge to turn my head to check his facial expression. "What?" I replied, in shock.

"Good shot," he repeated.

It felt surreal to have a fellow sniper confirm that I had grievously wounded someone. And he said it so calmly – like I had made a good putt on the green.

The fugitive and would-be cop-killer was prone now. He lifted his head for a moment and seemed to be looking around. I wondered if he was hiding his rifle under his body so he could ambush my team. I keyed up my radio headset with my left hand and told my teammates to stay back. "Threat is still active. Sierra is hot." I held the cross hairs over the base of his skull and waited for him to jump up and continue his murderous plan. Through my 10x Leupold Mark 4 scope, the 210-yard distance looked like twenty-one yards to me. From my perspective, I was hovering immediately above this killer's head with my rifle in hand. For the first time, I could feel my heart pound and see the reticle of my scope bounce in time with my heartbeats. I took a deep, slow breath.

The fugitive put his head back down and lay silently. There was a basketball-sized red stain seeping through his shirt. After a moment, one of the SWAT assaulters tossed a flash-bang near his head. The concussive blast rang out, but the fugitive didn't flinch. *No one can take a flash-bang at that range and not react*, I thought. *That guy's dead.* Moments later, one of the troopers fired a non-lethal bean bag round onto the subject and again he didn't move a millimeter. I keyed up my radio to let the team know I was putting my rifle on safe. "Sierra is cold."

The assaulters moved up and handcuffed the bomber's limp body. I was still floating virtually twenty-one yards above the scene through my 10x binoculars now. It felt like an out-of-body experience as I watched my two fellow medics, Dangler and Big Toe, work on the subject. Dangler put Quikclot combat gauze in a large wound where my crosshairs had previously been. Other operators started CPR with a bag-valve mask and chest compressions. An ambulance arrived soon but couldn't get the vehicle to the crisis site due to the terrain, so our operators loaded the subject onto a stretcher and ran him to the ambulance. I realize now that there was nothing even a surgeon could have done to save him. Even so, it was impressive how my teammates had endeavored to save the life of the man who had been intent upon killing them just minutes ago.

I was standing on the edge of the mesa now, just staring at the events. An Oklahoma Highway Patrol Cessna buzzed my position so low that I could see the rivets on the underside of the plane. The FBI plane was still circling quietly in its higher orbit. Mackey and I loaded up our equipment and prepared to crawl back down the south face of the mesa to my truck. Later that day, an Agent from the Evidence Response Team would have to make the same climb up this mesa to recover my cartridge casing for evidence, but he wouldn't be burdened by the weight we had carried. We dumped our gear in my truck and drove around to the north side of the mesa and found our teammates. Wiley gave me a big hug. There weren't many words spoken, but we all huddled around for a moment in silence. Later, Spence came over and told me to turn in my rifle for evidence. He said he had already collected carbines from my teammates who had fired as well.

The subject had had an AK-47 that was converted to full-auto, plus a magnum hunting rifle with a scope and several pistols. He had just been waiting for the moment to take out as many of us as he could. But the dog drew him out of the tent and slowed him down just enough for us to react before he could kill us. I found the Highway Patrol K9 handler who had risked his dog's life for us, and I gave him all the extra ammunition I had in my truck. His dog had saved lives that day.

AUTOPSY

After the autopsy was released, I learned that my bullet had struck the bomber mid-torso and torn through several organs in his abdomen. The temporary cavitation from the bullet had inflated a hollow cavity in his body large enough to stretch soft tissue to the point that it would tear. The expansion damaged his spine, which explained why he dropped so fast. The bullet had tumbled and broken into pieces as it travelled through his torso.

The Sierra MatchKing bullet I had used was made with one purpose: accuracy. The Open Tip (as opposed to Hollow Point) in the nose of the bullet wasn't to make it expand; it was because the manufacturing process sought to ensure that the boat tail (the end of the bullet) was perfectly concentric. If an imperfection caused it to wobble at the final millisecond that it left the bore, it would wobble during the entire flight. However, the unintended consequence of the bullet's jacket design meant that it would often break up when it hit a "soft" target, separating from its lead core with the various jagged pieces traveling different paths.

I had harvested several whitetail deer with this exact type of bullet and seen various results with my own eyes. Sometimes shreds of the bullet jacket had to be laboriously picked out of the meat, but sometimes the bullet punched a pencil hole through the animal. The bullet was capable of tremendous damage, but the fragmentation wasn't predictable. Luckily that day, it did its job.

The autopsy would show that in addition to my shredded bullet, two perfectly mushroomed, molybdenum-coated, bonded, FBI 5.56-mm carbine bullets were recovered from the lower torso of the subject. Those bullets most likely originated from Dangler or Wylie, who had had stable firing positions at the side of our vehicle, or from Twilight, who had been in the flanking position.

Dangler, who had just shot the man who was shooting at him, had then run up and tried valiantly to save that man's life. When the dust had settled, Dangler had had blood up to his elbows and soaked into the knee pads of his combat pants. He would have to wait for the medical report to find out if he had contracted AIDS or hepatitis from trying to save the life of the angry bomb-maker. He is the perfect example of the warrior ethos.

RESIDUE

I had had the opportunity to attend Lieutenant Colonel (LTC) Grossman's "Bulletproof Mind" seminar with my team in Tulsa months before the shootout at the Glass Mountains (also known as the Gloss Mountains). Afterwards, I read Grossman's excellent books, *On Killing* and *On Combat*. Subsequently, I had the honor of exchanging emails with LTC Grossman. He told me that random millisecond thoughts like, *"Is that an AK-47 or an SKS? Is that an EOTech attached? How did he attach it? There must be a picatinny rail on the dust cover,"* are not uncommon among warriors in combat and can be distracting. In my case, I had had enough preparation and training to prevent such distraction.

One theory is that the operation of the weapon is automated by rehearsal (motor memory); I had operated the safety and bolt so many times that the actions happened without conscious thought – much like how an athlete dribbles a ball or how people can drive home without thinking about the operation of the vehicle or where they are in the route. When an action is conducted "automatically," there is mental bandwidth available to think about other, often random, things.

In my correspondence with LTC Grossman, I also learned that, in stressful situations, there can be an instinct to engage in repetitive motion known as "perseveration." He said that there are documented instances of perseveration in which soldiers rack the bolts on their rifles and eject perfectly good ammunition. I also knew a former Air Marshal who admitted repeatedly decocking his SIG P229 pistol during high-stress drills, even though the gun was already decocked. The urge for repetitive activity creeps up when we are under stress. Because of my training, even though my subconscious brain told me to act, I had been able to resist that compulsion.

I also learned from reading LTC Grossman's work that a warrior should be *thanked* after a lethal incident. The words, "Good shot" from Mackey on Glass Mountain was the only encouragement I got from that op. And they meant the world to me. Everyone I spoke to afterwards engaged me with a sanitized air of indifference until the investigation was over.

And if his weapon must be confiscated for evidence, it should be replaced in that exact moment with another weapon. A faithful warrior must never be disarmed after a fight. The mental scar from disarming a warrior who has just seen bloodshed is tremendous. Even so, no one, including myself, had thought about what to do in the case of a sniper shooting. There are no spare sniper rifles. The idea of a spare is unreasonable. They are too expensive. The complete rifle with night scope was over $15,000.

It was painful for me to turn it in. I knew from my years on the Evidence Response Team that they needed to test my rifle for forensics and test the bullets for elemental make-up (percentage of lead, copper, etc.), so it could be matched to projectiles recovered from the deceased. They had to have my rifle, but I still felt a little like a child being disciplined as I turned over the rifle that I had trained with and bonded with for years. I wouldn't be able to train at sniper training or be available for sniper deployments for several months because my rifle would be in the lab in D.C. In hindsight, although each team couldn't afford to have a spare rifle, the boys at Quantico certainly had some spares in the vault. They gave my teammates replacement carbines the very next day but left me empty-handed. It would have meant the world to me if they could have told me a replacement rifle was in the mail from D.C., and I'd have been whole again.

I was in a quiet state of shock for the next week or so. Luckily, during our one-and-only group therapy session, the FBI lay-counsellor had told me to expect as much. I had dreams that replayed the events over and over again, but it wasn't scary or stressful. It was more like watching a movie on TV from the comfort of a sofa. Over the next year, I only told a few people about the shooting. I couldn't predict how people would react and I wanted to avoid unpleasant reactions. Eventually, my team would be awarded the FBI's prestigious Shield of Bravery. After that, it was hard to avoid questions, and I began to talk about it with some close friends and family. Even so, no one really understood. The most painful thing of all was that, even though I could talk to non-FBI friends, I was forbidden to talk about it with my teammates for almost a year until the local DA would provide a declination letter to the FBI. They didn't want us to accidentally commingle our testimony by talking to each other, but it was brutal to be cut off from my brothers when I needed them most. I understood that it was protocol to keep our testimonies independent, but it left a scar that never healed completely.

The bittersweet concoction of gloom and success after killing a man is difficult to deal with. You have spent countless hours sweating or freezing, enduring pain and boredom, so that you can be successful if you are ever called upon. The purpose of being a skilled sniper is to save lives by neutralizing someone who would otherwise be a murderer. Yet, when it happens, you are still overcome with the upsetting knowledge that you have separated the spirit from the body and left behind a mangled sack of meat. I've talked to other snipers and the feeling of, "Oh, man. I'm in trouble," was common in all of us, even when we were certain it was a righteous shooting.

Killing someone is appalling, bitter, and upsetting. Yet deliciously satisfying. We won the fight. I wasn't ashamed of it.

I met with a friend for coffee a few days after the gunfight at Glass Mountain. He had been a soldier deployed to Afghanistan and had assumed from various news articles that I was part of that sniper team. I told him about the burden on my heart, but also the deep satisfaction I had from saving my teammates from being shot.

He told me of a saying he had heard when he was in the Army, about "the man who carries the light bucket." If a man has been in a fight in which he must take someone's life, he will carry a bucket for the rest of his days. The man who reacts swiftly and decisively to the threat, but must end someone's life, will carry a light bucket. The light bucket comes from detesting the idea of killing but being forced to make a fatal decision. But the man who doesn't react, who doesn't train or sweat or push himself, who doesn't put in the effort and isn't ready when the decisive moment arrives – that man misses. He hesitates, he quits, he fails the team. That man carries the heavy bucket. My friend looked at me with a genuine smile, and said, "Be glad that you only have to carry the light bucket."

The ugly truth is that there is tremendous satisfaction in knowing that you were capable of doing what you had unrelentingly trained to do. For many of us, we have spent five, ten, even twenty years scrimmaging, training, testing, shooting, crawling, running, learning ballistic data, reading the wind, and knowing our dope. Finally, when the day comes, after hours of blinking away dirt, wishing you had some more water, laying on cactus and rocks that are your unwelcoming bed – one framed second in time presents itself to you, camouflaged by hours of sitting perfectly still and silent. In a flash, the moment comes and goes, and you know that your training was worth it. You are, all in the same instant, proud, relieved, worried, and uncomfortable that you killed a man – and overjoyed that you didn't let him kill your brothers.

Chapter 4

Operation Delta Blues

Mission #48
October 2011

"Is this the place?" I asked Lil Toe as we pulled up to a 50,000-square-foot indoor rodeo arena in Tunica, Mississippi. The small convoy of SUVs behind us contained our teammates.

"Hell if I know. This is the address, though," Lil Toe sighed. He wasn't in the best mood. He had been ferociously bitten by bed bugs last night in the hotel, and I didn't blame him for being grouchy.

That night we had settled into our rooms above a smoke-filled casino in Robinsonville, Mississippi. Only some of us got the optional bed bugs. Each of the twenty-five divisions of FBI SWAT teams that had been deployed for this operation had stayed in separate locations. Every location was far enough away from the arrest area that there was little chance of the 350 SWAT operators bumping into any of the seventy-two subjects we would be arresting later this week. This was Operation Delta Blues.

It was so named because it was in the region known as "the Delta," and the focus was cleaning up corrupt police officers. Hence the "Blues." Though the name was coincidental, the timing was perfect: the Blues Festival in the Mississippi Delta brought in 100,000 tourists, and it was happening the same week as this operation. That provided excellent cover for over 800 FBI SWAT and peripheral personnel to move into the area, settle in, and conduct reconnaissance on the criminal subjects. One of the case Agents, a local of the Delta himself, had sent an email the week prior, telling us to travel in less stereotypical "Fed" cars and clothes – preferably no black Suburbans, cargo pants, or shirts with American flags.

My teammates and I got out of our trucks and unloaded the large, wheeled bags that held our kits. A "kit" included body armor, helmet, tactical belt, rifle, radios and all the associated accoutrements. It didn't include our pistols, because we each had our pistols tucked under our shirts.

I pushed open a gray steel door that led into an enormous arena. The entire arena was one of two low-profile staging areas for the SWAT teams. The other teams were at the Memphis National Guard site. It smelled like the rodeos from my childhood – fresh hay and a trace of manure. There were waist-high tubular metal fences around the interior perimeter of the arena that defined twenty-five "pens." Each contained a couple dozen folding chairs and a few tables. They had posters above them to indicate the team for which they were designated. Teammates from other divisions were bustling around and unpacking their bags. About half-way down on the left, I spied an empty pen that read "OC" for Oklahoma City.

"Over this way, fellas," I called over my shoulder as I walked to our pen. Everyone pulled out their gear and set it out exactly the way they would want it, so that on the morning of the operation, when we jocked up, it would be ready to go. I set out my kit in reverse order so it would be simple to put on when I was up that morning, breaking through the fog of a sleepless night.

I leaned my armor against my metal folding chair after checking my communications box batteries, flashbangs, and infrared strobe batteries. I was designated to carry the shield on this op, so I removed the rifle magazines from the chest plate on my armor and put a holster for my back-up .45-caliber pistol in their place. Since I would be up front, I opted to wear more armor than usual and to carry two pistols instead of a rifle and a pistol. My body armor had Level-III Kevlar soft armor that wrapped around my torso. It would stop any pistol bullets but not rifle rounds. In addition, I added a Kevlar panel that dropped down over my groin. I had Level-IV ceramic plates in the front and rear of my armor that would stop just about everything, opting for the same to cover my shoulders. Having modern and modular armor available that allowed me to scale up or down in levels of protection was a huge improvement to the one-size-fits-none armor I had been issued when I first got on the team ten years ago.

I checked my headset batteries, plugged the headset into the comms box on my armor, and clipped the headset to a carabiner on the left shoulder of the armor so it wouldn't dangle. I clipped my gloves to my war belt, a wide belt that had loops to attach equipment. Mine had my pistol holster, extra magazines, handcuffs, and a medical kit. *Warbelt on the seat of the chair, boots in the center. Helmet under the metal chair. Uniform draped over the back. Shield leaning against the right side of the chair. Everything in its place.* One less thing to worry about on game day.

Over the next couple of days, I had a team leaders meeting planned as well as a medics meeting. An operation of this magnitude meant that we had to be prepared for casualties. We were dealing with neighborhoods so dangerous that local police had abandoned patrolling them. The possibility of corrupt district attorneys, judges and police officers added to the danger. All the medics were scheduled to meet to check med bags, rehearse medical scenarios, and review K9 emergency medical procedures. We took a walk through the helicopters parked behind the arena. Some were dedicated medical choppers and some were dedicated surveillance platforms. We had paramedics and doctors on standby for us, and veterinarians on standby for our furry warriors. A K9 handler gave us a briefing on how to treat a wounded war dog:

1. If the dog handler is injured, expect the dog to bite you.
2. If the dog is injured, you can do rescue breathing by holding its snout shut and breathing into the nose. Expect the dog to bite you.
3. For helicopter evacuation purposes: police dogs should be a lower priority than operators, but a higher priority than "bad guys."
4. If you load a dog onto a helicopter, restrain its legs first. Expect the dog to bite you.

GOOD OPERATORS MAKE GOOD CASES

Every operator on this mission was a Special Agent, and as such was a highly skilled investigator. I was currently overwhelmed with a multi-million-dollar embezzlement case at a large local machinery business. I was stressed about taking several days away from my case to complete this operation, but I was also relieved to change gears for a few days. SWAT operations were difficult and stressful, but they were so rewarding. When a long case started to drain my batteries, a good SWAT op would charge them back up.

For my embezzlement case I had issued a dozen new subpoenas to banks and websites, collecting evidence to prove that an insider had stolen millions from his own employer. That put me at forty subpoenas all together – forty printer boxes full of financial documents. I had a HQ financial analyst assigned to help me sort through it all, but she worked in an office an hour away from mine. Coordinating with her was difficult, especially since she had other cases competing for her attention. Absent any assistance, I collected and scheduled bank records and conducted surveillance and interviews on my own. That had meant twelve-hour workdays for me for over a year.

Since the victim in this case was a local business, it was important to me to make a solid case and support my community. Financial losses at local businesses meant people in my community losing income. I knew the prosecutive guidelines would dictate only about a twenty-four-month sentence for the subject, but real justice to me was taking back the stolen money. That meant I had to convince the local US Attorney's office to seize what the subject had bought with his ill-gotten proceeds. In this case, that included his luxurious paid-off house, a speed boat, a truck, two trailers, two tractors, and a cherry-picker crane.

Once my seizure warrant was approved, I would have to physically obtain these objects and transport them to the US Marshal's office an hour away. There they would be sold at auction and the proceeds returned to the victim. That's real justice. "Serve two years in a minimum-security federal prison and keep two million dollars when you get out" sounds like a game show people would willingly sign up for, not the consequence of a federal prosecution.

The seizure of all of the equipment was scheduled to take place when we got home from this op, but the federal agency that had agreed to help me backed out. In a panic, I sent a last-minute message to my teammates and asked them to help. Even though it wasn't their job, everyone answered me immediately and promised to make the three-hour round-trip drive to help me make the seizures once we got home.

That's the kind of guys my teammates were. We looked after each other when it came to operations or investigations – at home or on the streets.

DELTA BLUES CASE

The Delta Blues case had been underway for two years and involved six wiretaps, numerous informants, and hundreds of hours of investigation, surveillance, and analysis. It had started as a collaboration between the US Attorney's Office, the FBI, and the state police in an effort to address increasing complaints about police corruption. While it had uncovered criminal enterprise, drug trafficking, and gun violence, the core of it was crooked police, a crooked mayor, and allegations of crooked or incompetent judges and district attorneys. During the course of this investigation, six corrupt police officers had accepted bribes from drug dealers on five occasions each. The drug dealers had learned that they just needed to pay off the cops in order to rule the town.

Over the last several months, FBI surveillance teams had tracked all 72 subjects' "patterns of life" and knew exactly where 90% of them slept each night. But this would be the last night in their own beds.

The culmination of the case had been a long time coming. Something often misunderstood about the FBI is that, unlike state and local police, we don't make arrests early in the case. If a policeman makes an arrest, the subject is arraigned, the prosecution remanded to the state prosecutor, and the matter might not be heard in a state court for over a year. In the federal system, the arresting Agent works hand-in-hand with the local U.S. Attorney's Office until sentencing is done.

Under Federal law, a charged person has the right to a full jury trial within *70 days* of pressing charges. Once the "speedy trial clock" starts, the FBI can be compelled to provide all of its evidence quickly to provide the accused with a trial – even if the investigation isn't complete. A decent defense attorney will push for a speedy trial if there's even a hint that the FBI case isn't complete.

That's why the FBI arrests people at the end of the case, not the beginning. If we come for you, the case is already written up, surveillance completed, phones tapped, video footage uploaded. You've already been indicted by a federal grand jury, and you're going to jail for a long time. It can be disappointing from the outside perspective that the FBI doesn't make arrests at the outset, but the result would be criminals being released before all of the evidence can be collected. One of the largest crack cocaine dealers in this case had already been released because of a faulty case. This time, there would be no room for escape.

GPSs, SHARPIES, AND CHOPPERS

It was an impressive feat to organize an operation of this magnitude across numerous agencies and still maintain operational secrecy. One of the Case Agents had teamed up with another Agent named Nick from the FBI SWAT Operations Unit in CIRG (Critical Incident Response Group). Nick masterminded the staffing, funding, transportation, lodging, scheduling, and sequencing of this gargantuan operation. I had previously been a Sortie Generation Officer in the Air Force and been tasked with the logistics of multiple overseas deployments of my squadron of US Air Force jets and personnel. I knew that forecasting personnel and equipment needed for deployments was a monumental task with no room for error. I've never met Nick, but I can say with confidence that Nick's vision for this operation that had thousands of moving parts was executed masterfully.

Still, there were small unforeseeable complications. For example, several SWAT teams from the East Coast had been scheduled to fly in on military airlift. Unfortunately, the military agreement to support counter-narcotics missions had expired and the flight had been canceled at the last minute. These teams were scrubbed, but four other teams were able to be activated, with no notice. They had loaded up trucks with everything they needed and arrived with no time to spare to make sure we had all the operators required to complete the arrests.

Command and control personnel were working in a large FBI RV inside the arena. They already had sent all of the medical helicopter landing zones (HLZs) to my FBI-issued GPS along with every arrest location. Each team would be available to back up the next closest team if they had a critical incident such as a wounded operator or a gunfight. It was imperative to know what teams were operating close to you so that you could respond without delay. After being briefed on the area of operations, the HLZs, and the sectors of responsibility, I walked into the RV and flagged down an Electronics Technician from CIRG.

I extended my hand for a handshake. "Hey, brother. Sorry to bother you. Would you hook me up with the grid coordinates for the secondary and tertiary HLZ's for my sector? I'm lead medic for Oklahoma, and I want to make sure I cover all my bases."

He met my handshake with a smile. "Sure. We already programmed all the GPS's. You're good."

"Thanks, but I also want to put them in my personal GPS." I held out my GPS. I had purchased an upgraded version of the standard-issue GPS. "It doesn't take the same data card."

"Sure. I can put anything in anything, man. Let me see." He took the GPS, found a cable that fit it, and plugged it into his laptop. "There, you got it all. Every target, every HLZ."

"Perfect," I said. "Thanks!"

This technology seems elementary now, but at the time, it was revolutionary.

The GPS coordinates designated the HLZs, which were critical to the medical planning. Preparing for all of the medical contingencies for an operation of this size, while staying covert, was equally important. All the medics had convened the day before the operation and discussed plans for when, not if, an operator was wounded or killed. While we were together, we reviewed and rehearsed key medical procedures so that it would be fresh in our minds on the following day.

The Standard Operating Procedure was that all operators would write their designators (OC-2 in my case) with permanent markers on their left arms and not carry any ID cards. If there were corrupt police, they could have friends or relatives who worked at the hospital, potentially risking our identity and safety. We needed to be anonymous to all but our own personnel. We couldn't know for certain if someone on the hospital medical staff might be a friend or relative of the subject, or to what lengths they might go to exact revenge.

The local FBI office had an excellent relationship with a local ambulance service. Their medics were deputized by a local sheriff's office and could lawfully carry weapons to defend themselves. They were so well-regarded by the local FBI office that they were given authority to craft the medical plan in coordination with FBI CIRG. The intricate plan required that each twelve-man FBI SWAT team have at least one operator that was a nationally registered Emergency Medical Technician. That was Big Toe, Dangler and I on our team.

DANGLER

Dangler was an Army veteran, a fellow Christian, and a family man. He never took himself too seriously and always had us laughing, but he always pushed himself to improve. He had transferred to us from Kansas City Division. After an informal interview, fitness test, and firearms test, he joined our team. He was a fellow firearms instructor, medic, and sniper. He was average height and weight, and usually had a big, thick beard. He had a way of going on tirades about red tape and politics that was cathartic for himself and hysterical to everyone in earshot. He could rant about bureaucracy, lazy prosecutors, and poor leadership, and have us all belly laughing at the same time.

Dangler, (a.k.a. Gerlad) got his original call sign after he told the team about proof-checking his FBI credentials two months before he graduated from New Agent Training. To his dismay, his creds were printed with the first name "Gerlad" instead of "Gerald." He told the clerk who brought the creds for proof-checking that his name was spelled wrong. The clerk took a long look at his clipboard and said, "No. Sorry. It says 'Gerlad' right here. That's the correct spelling," and walked away. Gerlad stuck as his call sign after that…right up until he became Dangler, but that's another story.

In addition to our team's organic medics, there were advanced paramedics and trauma doctors from local hospitals and from FBI CIRG in strategic locations around the arrest areas. National Guard Counter-Narcotics helicopters (call sign "Mercy") were scheduled to be in orbit during the operation. Redundant HLZs were identified, named, and uploaded to every medic and team leaders' GPS. That planning would save at least one life on game day.

The day before the operation, Lil Toe went to a team leader meeting in one of the Skyboxes in the arena. This Skybox had the windows covered in newspapers because confidential sources were on site to give in-person intelligence briefings, including describing the layout of the subjects' houses. This gave us a distinct advantage. Once we knew the blueprint, we could tape off the measurements on the floor of the arena and conduct rehearsals in a "house" with invisible walls. It wasn't perfect, but it was one less thing to surprise us when the "execute" order came.

Besides the massive tactical Agent coordination at our locations, 150 additional support personnel from numerous FBI offices and state and local agencies were working behind the scenes, including six search teams and numerous interview and transportation teams. That was the backbone of the operation. It wouldn't do much good to serve all these warrants if there weren't staff to transport, interview, fingerprint, and collect evidence.

My team's first target was the residence of two armed and dangerous drug dealers. The plan was to drive up the driveway and from there, carrying a ballistic shield, I would lead the team to the front door. Frosty would be behind me with a ram. Mackey would drive another Suburban to the left side of the house in a blocking position in case someone tried to escape from the side door. The side door would also be our alternate breach point if the breach at the front door failed. While Frosty breached the front door, Dangler and Vulcher would remove an air conditioner unit from a window on the right of the front door, then stand on a small ladder, providing covering fire through the window.

It wasn't uncommon for us to "port" a window, meaning to break the glass, pull down the curtains from the inside, and provide a view into the structure. It typically involved one operator on his rifle and a breacher using a pike to break-and-rake the glass to prevent any cutting hazards. The port on this structure was complicated by the fact that it had an air-conditioner unit in it, and we had no idea how well it was installed. However, it was the only window on the front of the house and we needed to see into the room as soon as possible. Vulcher's assignment was unusual, but he was up to the task.

VULCHER

Vulcher had been selected for the SWAT team one year after entering the FBI. He was a former Border Patrol agent and a zealous Texas native. He had been on the team for about three years at this point and proved himself a solid operator.

His original call sign was "The Situation," an allusion to the show "Jersey Shore," because he frequently did his CrossFit workouts shirtless and was exquisitely man-scaped. His call sign changed when he told the team that he had been using Wylie's coveted Expedition while Wylie was deployed to Afghanistan. In an email to the entire team, he admitted that he "vulchered" Wylie's SUV and we made fun of him for the misspelling. Vulcher, however, embraced the call sign because he was quick to lose the "The Situation" moniker. The truth is, "The Situation" is five syllables and that didn't make for a terse call sign. The team was willing to go along with "Vulcher," but he had to keep the misspelling. It was arguably the coolest call sign on the team, and we'd be damned if we would let him spell it correctly now.

Vulcher and I had bonded over a lot of common ground. Like most of the guys, we were devoted husbands, fathers, and Christians. Additionally, we both grew up in financially lower-class families in small towns. Both of our fathers struggled with illegal drug use. My father moved out when I was two years old, but Vulcher removed himself from his own situation. During his 11th grade year, he moved in with his grandparents after his mother wrecked his truck after a drunken fight with his father. Shortly thereafter, his parents moved to Colorado without even telling him. His grandparents were his saving grace, and my mother and sister were mine. There's no doubt we both relished the mission of the team, but I think the reason we really loved the team was because it was *family*.

INFIL

In order to make it to the convention center in time, my team left the hotel at 12:45 a.m. Most of us laid in bed and rested, but none of us had really slept last night. Our orders were to be at the staging area to line up vehicles ninety minutes before departure: 1:30 a.m.

At the staging area, we kitted up in our designated corrals, then loaded up onto our SUV's. Waving flashlight wands, the logistics team formed up our 170-vehicle convoy in predetermined parallel lines. The teams were carefully sequenced so that the teams assigned to targets deepest into the town would be at the front of the convoy. That allowed all the vehicles to arrive at their targets at nearly the same time. Our convoy was set to roll out at 3 a.m. The north convoy was even further away and would roll at 2:30 a.m. The mission was set for 4 a.m., and our convoy movement would take 45 minutes to arrive at the final phase line.

As our convoy began to travel, six helicopters launched and a second 50-vehicle convoy was already rolling from the north staging area. The north convoy included one HRT team and four FBI field SWAT teams. Our convoy included State Patrol SWAT, HRT Team 2, Little Rock, Atlanta, Kansas City, Oklahoma City, Houston, and eight other FBI SWAT Teams. The armada had to be alarming for anyone who happened to be on the road that morning. One State patrol cruiser, three SUVs full of operators, one prison-transport sedan; another trooper, three more SUV's, another sedan – this pattern continued for miles down the highway.

After about an hour-long drive, each team arrived at Phase Line Yellow on time at 3:45 a.m. with one unified concern. Earlier everyone had loaded up with coffee, but hadn't seen a bathroom in almost three hours. Lil Toe was warning me that he was ready to pee on the floorboard. Since we were in a specified location in the convoy, it was not permissible to pull over for any reason. Now, at Phase Line Yellow, the entire convoy pulled to the shoulder and stopped. Lil Toe and I jumped out of the truck and waddled to the ditch on the side of the road. In unison, a ripple of troopers and operators spilled out of cars and trucks and jogged down the hill to the ditch. For as far as I could see in either direction, there were outlines of hundreds of men, backlit by headlights, peeing in unison. I started to laugh at the absurdity of it all, and I could hear laughter up and down the road. The scene was bizarre: there were hundreds of men, rifles slung to their backs, peeing into the night, and laughing in unison.

Lil Toe said, "Shadows and steam for as far as you can see."

I responded, "Phase Line Yellow, indeed."

Just then, an Army National Guard Blackhawk and an FBI Bell 412 slid past us at a slow, steady pace about 200 feet off the ground. With their lights blacked out, they were flying by night vision – with a full view of the spectacle of the mile-long pee-a-thon on the side of the highway. "I hope they haven't started video recording yet," I said.

My downlink flickered as the camera from NightStalker woke up and transmitted its FLIR surveillance feed from an altitude so high it was absolutely invisible from the ground. The image panned back, and I could see a wide-angle view of the entire town in the white-hot images from the thermal camera. I saw our first target in the feed. No one was walking around outside. The car in the driveway was black with a black-cold engine. It hadn't been driven in a while. *Let's hope they're asleep.*

FIRST WAVE

We mounted back up into the vehicles and waited for the clock to count down to 4 a.m. We had initially been set to execute simultaneous warrants at 5 a.m., but a command decision was made to move it up to 4 a.m. in order to mitigate potential hazards for school children departing for school. Just on the other side of Phase Line Green, the bridge at the end of the road, dozens of drug dealers and corrupt policemen were asleep in their own beds. It would be prison mattresses for them tomorrow night. Things were about to get exciting.

Two Agents had been assigned to knock on the police chief's front door at 3:55 a.m. to let him know we were arresting six of his officers in the next five minutes. The police chief had a right to know what was about to happen, but not with enough advance notice to affect the outcome of the operation.

Each twelve-man assault team had two additional interview/transport members with it. Negotiators, medical teams, radio operators, and search teams were on standby. Since armored vehicles were not yet commonly available to FBI SWAT, there were Agents in armored vehicles (known as "Rhino" teams) standing by to respond to a barricaded situation or an "Agent down" call. I checked my GPS and confirmed that the closest HLZ was the soccer field down the street from our first target. The guys counted on me to be on top of my game as a medic. If we had a casualty, I would get my teammate to the HLZ and stabilize him until the Mercy chopper got to us.

EXECUTE

The radio squawked in my headset. "All units, hold at Phase Line Yellow." Right on schedule, the TOC-Net radio crackled to life. "All teams, this is TOC. Compromise authority approved. All teams, execute, execute, execute."

Lil Toe echoed the command over the Fight-Net to our team, "Execute, execute, execute."

Our vehicle lunged forward. As I did before all high-risk missions, I recited Psalm 144 out loud.

"Praise be to the LORD my Rock, who trains my hands for war, my fingers for battle. He is my loving God and my fortress, my stronghold and my deliverer, my shield, in whom I take refuge."

"Amen," Vulcher and Lil Toe said.

Twenty-seven tactical teams, including two teams from Arkansas State Patrol, moved simultaneously to execute warrants on twenty-seven structures. Our team barreled down slushy mud roads. The neighborhood was run-down, and we passed several houses with boarded-up windows. Red and blue lights from the troopers who were blocking off the road flashed in my peripheral vision. Flash bangs detonated in the distance from another team's assault already underway.

It felt eerie to be an unwanted guest in a neighborhood that was so violent that local police had abandoned patrolling it. This was enemy territory.

But I wasn't alone. I felt the familiar warm wave of vigor, and I took a long full breath. I flipped down my night vision goggles as we turned into the drug dealers' driveway and saw the small white shack in shades of green.

"Mercy One is on station," came the voice of a medical evacuation helicopter over my headset as he announced that it had taken off and was in a holding orbit as planned. Mackey pulled his truck past us and parked on the left side of the house. I jumped out of our truck, pulled out my ballistic shield from the back seat, braced my left arm in it, and pulled it in close so that I had maximum field of view through the small bullet-proof window. I drew my .45-caliber pistol from my right hip with my right hand. My back-up .45-caliber pistol was firmly mounted on the front of my body armor. The fastest reload when you are shooting with a shield is to simply draw another weapon.

I led the team to the front door carrying the shield. Frosty was behind me with a ram and Lil Toe was behind him. Dangler and Vulcher set up a ladder at the base of the front window and crouched out of sight. Since I was closest to the door, I called out, "FBI. We have a warrant for your arrest. Come to the door now." There was a rustling in the house – we knew we were compromised.

Lil Toe ordered, "Hit it."

I stepped to the side of the door to make room for Frosty to swing his ram. It smashed into the door and fractured the locking mechanism but didn't open more than an inch.

At that same second, Vulcher and Dangler broke the mounts to the window air conditioner unit and pried it loose. Frosty took another mighty swing and the door snapped open about two inches, then snapped shut again. Vulcher leveraged the air conditioner upwards, anxious to get eyes into the house. However, rather than falling out of the window, it fell inside and *onto* the subject on the sofa, eliciting a shriek. That same sofa was blocking the door we were trying to ram open. Frosty hit the door a third time, but it wasn't budging any further. From the tiny crack in the door, I could see the sofa in our way.

Vulcher looked through the window where the air conditioner used to be and saw a gun by the subject on the sofa. The subject was jolted as the ram slammed into the door and rocked the sofa he had been sleeping on.

"GUN!" Vulcher called out. "Put your hands up!" But the subject ignored his commands.

The MedNet frequency blared over my headset from my long-range radio. "TOC, this is Norfolk. Shots fired. Agent down. Request immediate medevac."

It was 4:09 a.m. Nine minutes into the operation. *Well, we're off to one hell of a start.* Glorious serenades of flashbangs and explosive breaches were echoing all around us, surprisingly close. I could hear the ground and helicopter medical units responding on my radio. The UH-60 Blackhawk medevac helicopter from the National Guard called over the MedNet, "Mercy Air One, enroute to HLZ Turner," and left its holding pattern to rendezvous with the medical team dispatched to the wounded operator.

The Norfolk team would be out of the mission for the rest of the night. They would stay at their location while inspectors conducted a shooting review. TOC was already planning to shift Norfolk's second and third wave targets to other teams.

We might pick up a follow-on mission. This could be a long day.

Vulcher could see through the smashed window that the subject was stunned from the A/C unit falling on him. He bounced prone on the sofa with every strike of the ram, and that sofa wasn't going anywhere.

"Failed breach," I called out.

"Alternate breach," Lil Toe ordered. Since we had planned for just such a contingency, we moved to execute an alternate breach on the left side door of the house. Mackey was already there, and he positioned his rifle from behind the Suburban's engine block as we moved into position behind my shield.

DON'T RUN

In a flash, the side door flew open and one of the subjects bolted out the door, looking intently over his shoulder. Mackey and Frosty immediately sprinted at the runner and hit him with such force that he flew into the air and spent a millisecond hovering horizontally above their heads before gravity yanked him back down to the turf. I instinctively moved to block the doorway with my shield to protect them while they cuffed the subject. Another subject was standing in the room, just a few feet from the gun on the sofa. He glanced sideways at the gun, lifted his left hand and took a slow step to my right – toward the gun.

"Put your hands up! Do not reach for the gun," I ordered. I knew the shield I was carrying could protect my torso, but Frosty and Mackey were only partially protected by my shield. Besides being concerned for them, I tried to quash visions of a stray bullet hitting my exposed legs, severing my femoral artery and bleeding me out in 120 seconds.

My left hand was wrapped around the grip inside the massive Level-IV shield, and I lined up the sights of my .45 by looking through the bullet-proof window at the top of the shield.

He took another small shuffle toward the gun. I was in disbelief. *He's still going for the gun. Maybe he doesn't understand how this'll play out.* I decided to word it differently. I flipped on my .45's laser and made small circles on his chest. He looked angry and scared at the same time. I had seen big, crazy eyes like that before – from a feral cat caught in a trap. He took another step closer to the gun.

"Don't! I'm not kidding. I'll fucking kill you." My tone wasn't angry. It was sincere and matter-of-fact. He could grab that gun in under one second. I put my finger on the trigger of my .45. Vulcher's rifle was pointing through the window in the front of the house and my teammates were behind me, but no one had the shot but me. The subject and I locked eyes for a moment, then he suddenly seemed to understand he had only one choice: live or die. Both of his hands went fully into the air and the tension in his shoulders drained away. He didn't want to die today. Thank God.

"Turn around and walk backward to me," I said. A look of resignation fell over his face. And he did exactly as I told him. Once he was backed up to my shield, Frosty cuffed him and asked him, "Who else is in the house?"

"I don't know! Just my boy, I guess," he answered.

Since we knew the layout and had a solid perimeter, there was no reason to go rushing into a gunfight with whoever was still in the house. Time was on our side. We called out commands to surrender but got no response.

"Sending in the robot," Lil Toe said.

Lil Toe grabbed our small hand-held robot and the remote control from the backpack of another operator. The throw-bot was the size of a toilet paper roll with donut-sized tires on the ends. He pulled the pin that activated it, like pulling a pin on a grenade, then lobbed it over the top of my shield. It bounced off the wall and toppled sideways into the hallway to the left, automatically righting itself. Everyone stood behind my shield as the little "throwbot" started clearing the house. Lil Toe drove it through the two bedrooms, but didn't see another subject.

"Nothing seen. Manual clear." Lil Toe said. He put the robot's remote control in a dump pouch on his belt.

"Copy that. Bang it again," I said. I was looking forward to putting the 65-pound shield down.

"Bang out," Frosty said as he bankshot a flashbang grenade that bounced off the same wall as the "throwbot." Before it detonated, we were already moving swiftly through the house. We cleared both rooms, looked under beds, in showers, in closets. Nothing. *Maybe this guy didn't know that his buddy already tried to escape out back.*

"House clear," Lil Toe called out after we had searched every space in which a human could hide. "Where's the other guy?!" Lil Toe asked. I looked at Lil Toe. "I think the 'other guy' was his buddy who squirted out the side door. I think it's just the two of them."

We marked and unloaded the subject's pistol for the Evidence Response Team and turned over both of our subjects to the transport team. As we collected our ram, shield, robot, and whatever we may have dropped along the way, an Arkansas State Patrol trooper ran the names of our arrestees. Neither of them were the drug dealers we were initially looking for, but they both had outstanding felony warrants. While they weren't the guys we had expected, they both needed to go to jail.

Lil Toe keyed up his TOC-Net radio. "TOC, this is OC-1. Target Alpha Seven is clear. No injuries. We did not locate subjects 14 or 15, but subjects 22 and 23 were at location Alpha Seven. Both are in custody and turned over to the transport team. Let ERT know the structure is safe and ready for search."

"OC-1, Standby," the TOC answered.

Another voice crackled on the radio, "TOC, HRT Team One, be advised, we took fire at Mariana. No injuries. Subject surrendered without further incident."

"Copy, HRT One." The TOC was having a busy night.

"OC-1, this is TOC. You are cleared to move to your second objective."

Lil Toe pointed his index finger in the air and swirled it in a circle. "Mount up!" We finished loading our gear and climbed back into our trucks.

As we drove to the second target, I thought about the barrage of noise I had processed in the last few minutes: both MedNet and Fight-Net wailing on my headset, not to mention the verbal commands and flashbangs echoing in all directions. It had been an assault on the senses.

****

THE BLUE ZOO

I flashed back to my freshman, or "fourth classman," year at the U.S. Air Force Academy. It was early in my second semester. It was cold and miserable this time of year in Colorado Springs, and we referred to it as the "dark ages." I had figured correctly that the six weeks of Basic Training was going to be hard, but I had no idea that the same intensity and treatment would continue the entire first year. It was like year-long Basic Training, except now we had Ivy League-level college classes, intramurals, room and uniform inspections, and fitness tests piled on. It didn't help knowing that the Air Force brought in 1,500 freshmen each year with the intention of only graduating 1,000. They were more than happy to start kicking out one-third of each class. If you were a "four degree," no one was rooting for you.

"In the hall, SMACKs!" (Soldier Minus Abilities, Coordination, and Knowledge.) Some sophomores, known as "third classmen" or "three degrees," had called us into the dormitory hallway for evening "training" (a nice word for hazing). We were all still in Class-A uniforms and we had pulled another all-nighter, as usual, cleaning for the SAMI (Saturday A.M. Inspection) or "Sammy." Our squadron had lost some points because one of the urinals was wet during inspection. It was clean, but it hadn't been perfectly hand-dried. I had the unfortunate duty of urinal cleaning for this SAMI…again. I had done a good job cleaning and drying them, but if an upperclassman urinated in one right before the SAMI, it was a freshman's job to clean it again.

"Give me fifty, Rebmann!"

I dropped into the front-leaning rest position and started calling out my pushup count.

"One! Two! Three…!" My classmates dropped and joined me in solidarity while three-degrees began to select individual fourth-classmen for special "training." We were going on two days without sleep because it was a "triple threat" weekend. That meant a parade, an in-ranks inspection, a mandatory football game in Class-A uniform, *and* cleaning all night for a SAMI. There was little time to prepare for my Physics exam Monday morning.

"Schofield's quote! Go!" A three-degree had commanded me to start reciting one of the many passages we memorized from our *Contrails* knowledge book.

"Sir, Major General John M. Schofield's graduation address to the graduating class of 1879 at West Point is as follows: 'The discipline which makes the soldiers of a free country reliable in battle is not to be gained by harsh or tyrannical treatment. On the contrary…'"

"How many push-ups is that, Rebmann?" I heard in my left ear. A three-degree was two inches from my face. I had forgotten to keep count. I responded with one of the "seven basic responses" that we were allowed to speak during this twelve-month marathon as "unrecognized" cadets.

"Sir, I will find out," I replied. We weren't allowed to simply say 'I do not know.'

"Why the hell don't you know, Rebmann!?"

There was only one authorized response of the seven I had to choose from. "No excuse, sir."

In my right ear: "He didn't fucking say 'stop,' Rebmann!" I heard the voice from another equally angry three-degree that I didn't dare look at. I kept my eyes "caged" straight ahead at the wall across from me. The cacophony of upperclassmen yelling at four-degrees rippled up and down the hallway and made it hard to hear commands unless you really focused. There was no one here but cadets. No officers to oversee. I fought back a recurring seed of doubt whether six more months of this, in order to be an Air Force officer and pilot, would be worth it. Or whether I could hack it.

I resumed my push-ups while I restarted the quote and took mental note of my push-up count.

One.

I shouted at the top of my lungs, because the only appropriate volume when you were being "trained" was maximum.

"Sir, Major General John M. Schofield's graduation address to the graduating class…"

Two.

"…of 1879 at West Point is as follows: 'The discipline which makes the soldiers of a free country…'"

Three.

"'…reliable in battle is not to be gained with harsh or tyrannical treatment…'"

Four. What an ironic quote.

"'…On the contrary, such treatment is far more likely to destroy than to make an army…'"

Five.

"Stand up! Knees up! Now!" rang in my right ear. He held his hand palm-down at waist height, and I knew what that meant. I jumped to my feet and ran in place. Each of my knees alternated as they struck his open hand. His hand rose until I was jumping to make contact with it.

A shriek in my left ear, "He didn't fucking say you could st—"

I caught myself pausing, and continued to shout, "'…It is possible to impart instruction and to give commands in such a manner and in such a tone of voice as to inspire in the soldier…'"

"Better." I heard a new voice in my left ear. A calm voice. A first-classman's voice.

I dared not pause. "'…no feeling but an intense desire to obey, while the opposite manner and tone of voice cannot fail to excite strong resentment…'"

He spoke directly into my left ear. "Rebmann, you want to be a pilot? Do you think you can communicate on the radio and listen to your wingman while your radar warning system squeals? While your heat seeker growls? And still maintain your altitude and throttle settings while you're in a fight for your life?

I knew better than to respond before I finished the quote.

Right ear: "Knees up!!!"

Right, left, right, left, right.... I automated my legs in tempo with my voice.

"'...and a desire to disobey. The one mode or other of dealing with subordinates springs from a corresponding spirit in the breast of the commander.'"

Right ear: "Push-ups. Go!!"

I dropped without stopping the quote.

One.

Calmly, in my left ear: "Seems like you'd have to sort through a lot of chaos and still control your body and mind, doesn't it?"

Why hasn't anyone explained it to me like that before?

Two.

"'He who feels the respect which is due others cannot fail to inspire in them respect for himself while he who feels, and hence manifests, disrespect towards others...'"

Three.

"'...especially his subordinates, cannot fail to inspire hatred against himself.'"

Four.

I slipped off into a dissociative state, as I had learned to do. I could see myself doing pushups as if I were a camera in the ceiling. It was easier like this – looking down on myself from above. Maybe this wasn't *just* hazing. Maybe there was some point to it. *Separate the inputs from your right ear and left ear. Separate channels. Automate your actions. Separate what you hear from what you say and what you do with your body.*

****

SECOND WAVE

Our second target was an armed and dangerous drug dealer who lived in a fleabag motel room. As planned, we stacked up on his motel room door and prepared to breach and gas him out. We had contacted the on-site motel manager and requested a key to the room. Surprisingly, the manager refused, so our alternate plan was to breach the door or window if he didn't come out.

As we exited our vehicles, distant flash bangs and breaching charges echoed to the constant drumming of medical and surveillance helicopters overhead.

Wiley looked at Frosty. "Choppers that low and blacked out remind you of Afghanistan?" he asked.

Roughly half of our team had volunteered to serve in Iraq and Afghanistan in numerous roles, including prisoner interviews, managing drone feeds, and going downrange with military units like SEAL teams and the 75th Ranger Regiment. Big Toe, Frosty, White, Gobbler, Wiley, and Jimbo had all volunteered for two- to-four-month deployments.

Putting FBI operators on the ground, who could immediately collect physical and cyber evidence, identify deceased combatants via DNA, conduct direct interviews with combat detainees, and corroborate military drone footage, significantly enhanced the entire US counterterrorism program. The FBI had developed a great reputation because of the expertise of our interviewing skills. There was no water boarding, torture, or threats. Our technique was to build trust and determine deep motives – what made people tick.

The core mission of the FBI is to determine the truth, and the key skill set of an FBI Special Agent is to identify deception and develop a rapport so that people feel comfortable and motivated to tell the truth. Convictions aren't the goal, getting the truth is. Proving someone innocent is as much of a "win" as proving someone guilty.

I led the team up the rusty metal stairs to the subject's second-floor room and held my shield so that it would protect Frosty and Lil Toe behind me. Mackey and Twilight slipped silently past the door and window and stood out of sight on the left side. If Frosty couldn't breach the sturdy steel door with his ram, we weren't going to stand on the balcony and take fire. The tertiary plan was for Twilight to smash in the window and deploy gas or bangs.

Everyone was in place. I kicked the door twice with my boot. "This is the FBI. Demetrious Jones, we have a warrant for your arrest. Come out with your hands up!" I called out.

A few seconds passed and there was clattering in the room.

"He's up," I said.

"Prepare to breach." Lil Toe was giving Frosty and me a second to plan our dance moves because I would be shifting over for Frosty to swing his ram.

The door cracked open and clattered on the safety chain.

"Contact!" I called out. "Open the door the rest of the way. Now!"

The door shut, the chain clattered, and then the door opened again. Frosty dropped his ram and drew his pistol.

"What the hell?" the man asked.

"Demetrious Jones?" Frosty asked him.

"Yeah," he answered.

"Put your hands behind your back. You're under arrest." Frosty explained.

As the cuffs clacked closed on his wrists, I scanned the inside of the room. There was a semi-automatic pistol on the stand right beside the door. We were all lucky that he hadn't decided to fight it out tonight.

FOLLOW ON

With the gun unloaded and marked for the Evidence Response Team, the subject was turned over to our second transport team. We collected our gear and returned to the trucks again, while Lil Toe called the TOC on his cell phone to confirm arresting Subject #55.

Lil Toe called over the radio for everyone to rally on his position. He pulled off his headset and wiped his forehead with the back of his glove. "Guys, we have a no-notice follow-on. It's a search warrant. No subject's expected. We have zero intel. This was probably a last-minute warrant, so we don't know the layout. There hasn't been any surveillance. We don't know if there are dogs or kids or hazards."

When we got to the house, we got off the truck and assembled in the same order as before and assumed the same roles. Frosty, Lil Toe, and I were moving toward the front door of the house while Twilight, Mackey and the other guys were setting a perimeter. The sun was coming up and giving me a little burst of energy. We had been up all night and were running on caffeine and adrenaline. As we closed in on the front door, I heard Mackey's voice over the radio.

"Contact! White-red corner."

I kept my eyes forward. As the shield man, I didn't have the luxury of looking over my shoulder to see what was happening. I had to face forward and protect the team from threats from the house. In my peripheral vision, I could see a man getting cuffed by teammates.

"Subject secured," Mackey announced.

We knocked and announced at the front door, but no one responded. Frosty breached the door and we conducted a slow, methodical clear. The house was empty.

Lil Toe called TOC. "TOC, OC-1. Follow-on location cleared."

"Copy, OC-1. Hold that location until a search team arrives. Then return to staging and stand-by for additional follow-ons."

The adrenaline was starting to wear off. The weight of the extra body armor and the shield had sapped my energy a little faster than usual. Twilight and I walked out into the yard and sat down side-by-side on a pair of children's Big Wheel tricycles. I leaned back against the chain-link fence and relaxed my back. Twilight took out his cell phone and took a selfie of us. It was ridiculous: the two of us sitting on children's toys wearing our full armor. The silly photo and the moment of mental decompression caused us both to start laughing. I hoped no one saw us, because we looked like crazy people.

"What was the 'contact' as we were on approach?" I asked Twilight.

"Dude!" he answered, "This guy just walked up and said, 'Hey, I hear you guys are looking for me.' So, Mackey and I cuff him up, check his ID, and he's Subject #4!"

"What? That's nuts. He just walked up and surrendered?" I asked.

"Yeah. He said his phone was blowing up. He heard everyone was getting arrested today and he just wanted to get it over with." We both smiled.

RESULTS

The search team pulled up to the house, so we turned over the scene and headed back to the arena to sit on standby for any other follow-on warrants. Back in our corral, we each started to re-organize our gear, load extra bangs and water, and prepare to go out again. There was talk of follow-on missions. Despite our exhaustion, if one was available, we didn't want to miss it for lack of being prepared. Frosty leaned his ram against the wall and sat in his metal chair.

"Hey, Frosty," I said.

"Yeah, Money," he answered. His eyes were closed, his helmet in his lap.

"Were you holding the ram backwards or something?" I walked over to grab a couple bottles of water.

His eyes opened. "What!?"

"It just seemed like the more you hit the door on the first target, the more it stayed closed. Maybe you were holding the ram backwards," I said.

"There was a sofa in front of the door! What the hell do you expect?" Frosty snapped.

I handed him one of the bottles of water and smiled. "How hard can it be to open a door? Isn't that literally what breachers do?"

As I walked away, shaking my head, I heard Frosty laughing as his voice echoed in the arena. "You're a dick, Rebmann."

Without looking back, I held my hands up in false despair, and replied, "Just…you know…open the door."

I walked over to Lil Toe. He was pacing around in the middle of the arena by himself. I think he just wanted a minute away from everyone to reset. He was rubbing his temples with his fingertips. I could tell he was mentally exhausted. We had been the ones effecting the assaults, but he had had the burden of navigating to each location, coordinating resources, and constantly keeping a headcount. I knew he was in no mood for joking. "Any more follow-ons?" I asked.

"Nope. They're cutting us loose. Tell the guys to get ready to head back to the hotel. Let's get some sleep and then we'll all go out for dinner at six."

I looked at my watch. It was just past noon. "We're done?!"

"Yeah." Lil Toe replied. "Tell the guys to start packing up. We got sixty-eight of 'em. There was a handful we didn't get, but they're listed as fugitives now. We'll get 'em all."

I was so proud of how the mission had gone. Not just how well my team worked together, but the way the FBI and all of the other agencies worked in unison. I was slightly sad that it was over, but I was also ready to get showered up, nap, and go hang out with the guys.

In order to ensure 68 initial appearances in federal court, the National Guard was brought in to coordinate buses to transport the subjects. The local US Marshal's Office stepped up to provide prisoner security and courtroom security. Facts uncovered after the arrests indicated the drugs in these towns were being supplied from Mexican cartels. One of the Delta Blues case Agents traveled to Mexico with a plan to arrest the woman that was masterminding the drug trafficking. Unfortunately, Mexican Immigration refused to remand her to US custody because she had a young child.

That last night in Mississippi, there was just a quick clip on the local evening news about the raid. Several news sources didn't carry the story at all. Even so, the operation was a tremendous success. It took thousands of man-hours and hundreds of thousands of dollars, but we made a significant contribution to the welfare and safety of the citizens in the Mississippi delta. This, like many national FBI SWAT operations, doesn't always make the news, but it does make a difference.

In addition to the successful execution of an enormous arrest sweep with no loss of life, we had seized $470,000 in US currency, 39 guns, and 3295.8 grams of drugs. There had been two shootings, and one FBI operator was shot in the leg, but he would recover. No subjects were shot or injured.

No one even got bit by a dog.

Chapter 5
Wewoka
Mission #62
July 2013

My phone vibrated persistently on my nightstand. Part of me hoped that it was a callout and part of me wished that it was just a wrong number. I rubbed my eyes and flipped over the phone and saw that it was the FBI Communications Center calling. My alarm clock flickered from 11:29 p.m. to 11:30 p.m. I hadn't been asleep for even an hour.

"Henloo?" The word didn't come out right. For no good reason, I tried to make it sound like I wasn't just startled awake.

"Special Agent Rebmann, there has been a SWAT call-out. The rally point location has been sent to your phone."

"Got it. Thank you. I'll be en route right away."

I tried to quietly slip out of bed, but I knocked my phone off the nightstand and it clattered on the floor. My wife, ripped out of a dream, looked up at me. Then she closed her eyes and pulled the blanket over her head. She knew the drill.

I pulled up to the rally point in a public parking lot in tribal trust land near Wewoka, Oklahoma and pounded the rest of my energy drink. In the pitch-black parking lot, we formed our standard circle. I fought back a yawn as Lil Toe began the mission brief.

"Our subject is a member of the Seminole Nation…"

One of the lesser-known facts about the FBI is that it is responsible for investigating felony crimes in "Indian Country," which is how it is listed in Title 18 of US Code. We work with tribal police, if there are any, and with the US Bureau of Indian Affairs, but ultimately, major felony crimes on tribal land are the responsibility of the FBI. In my career, I had investigated assault, rape, embezzlement, and murder for the numerous tribes, including Kaw Nation, Sac and Fox, Absentee Shawnee, Kickapoo, Seminole, Choctaw, Citizens Band Potawatomi, and Otoe-Missouria.

"He has been charged with violent sexual assault, and a federal warrant for his arrest has been issued," Lil Toe continued. "He was located in a small trailer in an open rural acreage. We will send in a sniper team to observe until sunrise and then effect a call out." We didn't conduct warrants before sunrise unless it was absolutely critical. We defaulted to conducting "knock and announce" warrants after dawn to prevent any unnecessary confusion for the inhabitants of the residence. It was a little more dangerous, but it was worth it to give the resident every chance to surrender and to prevent anyone from thinking it was a home invasion and shooting at us while thinking they were fighting off robbers.

I tucked my thumbs under my armor and pulled it away from my chest to help me stay cool. Even in the middle of the night, it was hot and sticky.

"'Agents may use deadly force only when necessary, that is'..." Lil Toe read the FBI Lethal Force Policy out loud even though everyone had it memorized since New Agent Training. We had all heard it recited hundreds of times, but there was always at least one reading of it before we deployed as a final reminder.

FBI SWAT had conducted countless high-risk arrests over its 50+ years of operation, and there had been a substantial number of gunfights, but not one gunfight in which the use of force wasn't absolutely necessary. And there hadn't been a single excessive force complaint. The FBI Lethal Force Policy gave us a wide berth to fight to protect ourselves and others, but we prided ourselves on always using restraint.

There's probably no better example than the instances in which we had been shot at and the subject had dropped his gun after running out of ammunition. In every instance, we had held fire when the subject was no longer armed. That takes a gargantuan amount of self-control. I would argue that it takes more discipline and moral courage to *not* shoot after you have been shot at than it does to run into a room full of armed drug dealers.

The team was always tight, but our *esprit de corps* was as high as it had ever been. We had been deployed to San Juan for Operation Main Hub just a few months prior. There, we had been part of a 500-operator mission to arrest drug lords. We were the first of forty twelve-man SWAT teams to enter the compound and deal with armed thugs in guard towers around the complex. We had been escorted by five helicopters and two fixed wing aircraft. It was a tremendously successful mission. In the fourteen hours between 2 a.m. and 4 p.m., we had arrested 126 criminals. We then celebrated the night after the mission by getting to see the town and experiencing the excellent food and drink. Well…mostly drink.

Once we broke the circle, I walked back to my truck and started checking my sniper gear. I was sweating through my uniform. The air was a little cooler since the sun had gone down, but it was still a typical July Oklahoma evening – muggy and 85 degrees with no breeze.

Dangler was two cars down from me and was pulling his gear out as well. We were scheduled to go to a gucci long-range sniper school in south Texas together for a week and we were excited to hang out on that trip.

"Dangler! You ready to get torn up in briar patches?"

"You know it, Money," he said. He had a spring in his step.

Dangler was a solid gunfighter, not to mention a solid guy. I loved doing overwatch operations with him. He had earned his new call sign after he fell through the ceiling of a drug dealer's house in Arkansas while searching the attic with Vulcher. He caught himself on a rafter as he fell, and Vulcher pulled him back up. I was standing in the hallway under the attic, so I had the surprise of seeing Dangler's legs erupt from the ceiling and dangle in front of my face. After that, he was forever the Tenacious Dangler.

Shallow, a.k.a. Mean Gene, a fellow sniper, came over to Dangler and me. "I wish I was going out with you guys."

Shallow worked in a rural RA in southern Oklahoma. He and I were both fans of Pontiac muscle cars – specifically Firebirds. He actually owned a mint black and gold Pontiac Trans Am that had been driven by Burt Reynolds in the movie *Smokey and the Bandit*. Shallow's FBI partner in his RA was a small-statured black man who had been an attorney before becoming an FBI Special Agent. His name, of all things, was Burt Reynolds. You can't make these things up.

"I wish you were too, Gene. At least you get to catch a long nap with the team while we're out in the woods for the next six hours." Gene had been assigned to the assault team for this op, despite his preference to deploy as a sniper.

"Nah, man. I'm all amped up. There's no way I'm gonna be able to sleep." He shook my hand and wished us luck.

A supervisor standing next to an ASAC saw Dangler and me in multicam camouflage uniforms and said, in a singing voice, "One of these does *not* look like the other…" and chuckled to himself as he sipped on his gas station cappuccino. Dangler and I stood out since the rest of the team was wearing olive-drab uniforms, known as OD green. We might have worn OD as well, but we would need all the camouflage we could get. The supervisor thought he was being clever.

"Forget which outfit to wear?" he giggled.

As I walked up to him, I made my most sincere face. "Thanks for being out here, Bill. Seriously. We're all grateful for what you and all the other supervisors do to help during high-risk ops."

He hesitated. "What do you mean? What do I do to help?"

"Yeah." I patted him on the shoulder. "Exactly." And I walked away. There were legitimate reasons that some supervisors didn't like me, and I was okay with that.

As I walked over to the parking area, I saw a couple of breachers setting up cots and climbing into the backseats of their cars to sleep.

"Nighty night, sweet little eight-pound six-ounce baby breachers. Sweet dreams. The snipers will take it from here." I said jokingly.

"Call us when there's a door to kick in, sniper divas. We all have our roles," one of them answered.

"Of course, of course," I said seriously. "Some of us have to know how to stalk and shoot, navigate at night, patrol under NODs, and conduct clandestine reconnaissance. Then others of us have to, you know, just…open the door. We all have our roles." I grinned at him.

"Go take your $10,000 rifle and enjoy your camping trip, Money. You know you love it." He wasn't wrong.

"See you boys in six hours. Be safe," I said. I looked at Dangler who was laughing at the false rivalry. He said, "Let's do this, Money," and he flipped down his NODs.

MUD SLIDE

Dangler and I trekked through terrible thickets to get to our sniper hides where we could observe the subject's trailer and be in position before the sun came up. We cut several barbed wire fences with our multitools and climbed in and out of muddy ravines where my feet sank into sloppy mud with each step. I managed my gait in order to pull my feet straight up after each step, reducing the distinctive sloshing sound that humans make when they walk through mud. Ahead of us, there was a ravine about forty-five feet across and of uncertain depth. We were too close to the roadway to risk using red lights, much less white lights, so all I had was my NODs to estimate the size of the ravine. It extended for miles to our right according to topographical maps we had reviewed, and it extended to the paved roadway to our left and passed under a bridge. We had no choice but to move straight ahead and through it.

"That looks pretty deep," I said to Dangler. "I can't see the bottom."

"Yeah, but we're good," he answered in his standard optimistic way.

"I'm not worried about getting in, I'm worried about getting out," I said.

In order to free up both hands, I slung my M4 carbine over my back where my bolt action rifle was already resting. I scanned the ravine again with my NODs, but there just wasn't enough ambient light to see clearly. It looked like an infinite black hole. I reached up with my left hand and clicked the switch on the infrared light on the side of my helmet. It gave me a little more depth perception, but I still couldn't see the bottom.

Overconfident, I squatted down to lower my center of gravity, leaned over the edge of the ravine, and eased my left foot forward. I could feel the mud slide under my right foot, and I poured over the edge of the creek's bank. My whole body slid forward, and I shot down the edge of the ravine like I was on a playground slide made of mud.

I instinctively held my legs together as they had taught us in free-fall parachute training. If you were parachuting into a forest, you press your legs together as tight as you can. It will strengthen your landing, and equally importantly, if there's an errant tree branch in your path, you do not want it to find its way between your legs.

Glop! My feet hit the bottom of the ravine and my boots sunk a few inches into the saturated soil.

"You good?" Dangler whispered.

"Come on in. The water's fine," I answered.

He slid down behind me in an equally ungraceful manner. We did our best to step on stones getting across the creek to keep from falling in, and I was thrilled that I hadn't gotten water in my boots. On the far side of the ravine, Dangler and I looked up at the embankment.

We were fifteen feet below the edge. Even if we stood on each other's shoulders, we couldn't reach the top.

Through my NODs I saw Dangler gesture to a root sticking out of the creek's edge about six feet up. I knelt down on my right knee and made a 90-degree angle with my left knee. I looked at Dangler, gave a nod, and patted the thigh of my left leg. He gingerly set his right boot on my thigh and used it as a step to reach the tree root. He grabbed it tight and pulled himself up until his hands were in the center of his chest.

Here's a good example of why we have to pass the pull up test while wearing armor and a helmet.

I stood up and put one hand under each of his boots and pushed him up until he could grab another root. He swung his weight back and forth as he made his way up. I stepped to the right to begin my climb. If he came tumbling down, I didn't want to be in the way. I kicked forward with the toe of my right boot and it sank into the muddy riverbank to make my own step. I got a little height before it started to break loose, so I kicked a hole in the bank with my left foot and luckily found a good purchase for my boot. I grabbed the tree root above me and pulled myself up the way Dangler had, but I couldn't find another root.

I was surprised how hard I was breathing. *This isn't that hard. I shouldn't be this smoked.* Some simple activities are substantially harder to accomplish when your body needs sleep, and I was starting to feel it.

I held myself in a pull-up position and tried again to find a root. *This isn't gonna work.* I was about to ease myself back down the riverbank when I saw Dangler's boot above my head. *Crap. I must not have moved out of his way.* His boot wiggled just above me.

Then I realized he had gotten to the top, bear-hugged the base of an elm tree, and was extending his leg down so I could grab it. I didn't hesitate. I climbed him like a monkey going up a rope ladder. I grabbed his ankle, pulled myself to his belt, then used his shoulder to pull myself until I could get my hand above the ravine.

We both got to our feet on the edge of the bank and checked to make sure we still had all of our gear. I started to laugh to myself a little. My gloves and boots were caked in mud, and the knees and seat of my pants were saturated in wet red clay.

"What do you think the assaulters are doing right now, Dangler?" I said.

"Sleeping. Those bastards are sleeping right now," he answered.

My headset came to life. "Sierra One, Voodoo, how copy?" I heard Big Al's voice in my headset and instinctively looked up at the night sky. With my NODs on, his strobes were easy to find.

"Solid copy, Voodoo. How me?" I answered.

"You're 10-2, Sierra One. I lost your beacons for a while. You guys good?"

"Roger that, Voodoo. We've been playing in the mud. We're good to go. Thanks for checking on us."

It was time to check in with Lil Toe. I was secretly happy that someone else had to stay up all night with us. "OC-1, Sierra One." I said. There was no reply. I tried my second radio, but it didn't get through either.

"Voodoo, Sierra One. Can you hear my transmissions?" I asked Big Al.

"That's affirmative, Sierra One," Al answered.

"Will you relay to OC-1 that we are passing checkpoint Echo?" I asked.

Since Big Al was above us, the trees weren't blocking our transmission to him, and since he was at altitude, his signal could reach back to the rally point without interference. *It's always good to have Voodoo up on an op.*

I activated the infrared laser on my rifle and made circles with it at our intended direction of travel. "Voodoo, Sierra One. Do you see my lasso? The objective is 300 meters in that direction. Can you confirm?"

"Negative, Sierra One." Big Al answered.

"Can you light it up?" I asked.

"Stand by. Arming laser now."

I looked through my NODs in the general direction we intended to travel. I couldn't see more than ten yards in any direction because of the forest, but Voodoo "lasing" the target would still be helpful. For just one second, a laser invisible to the naked eye zapped from a small blip in the sky and landed somewhere just past the trees in front of me. It was helpful confirmation that we were on track despite playing slip-and-slide in muddy ravines en route.

"Perfect. Thank you, Voodoo."

SNIPER'S HIDE

I low-crawled out of the tree line towards a large cedar tree that I planned to make my home for the next several hours. Dangler and I split up. He was going the long way around so that he could see the rear and left side of the trailer.

I settled under the cedar tree in a position where I could see the front and right side of the subject's trailer. I literally climbed into the middle of the enormous tree so that the branches, which extended to the ground, were natural camouflage. I was terribly allergic to cedars, but this location was ideal.

I took off my NODs and helmet, and put them in my ruck. I pulled out a ghillie hat with strings of jute across the face to make sure the outline of my face was broken up. I put my M4 carbine in my ruck and pulled out my bolt action precision rifle. The prairie grass was too high for me to see over if I were in the prone position. So, I extended the legs of a tripod and locked my rifle into it so that it was a couple feet off the ground. Then I sat cross-legged behind it, pressed the stock into my shoulder and got good eye relief behind the scope. It was a pretty decent position. I could sit like this for a while before my legs started going to sleep. I was invisible in this tree, and I had a great view of the objective.

I slowly tugged at the Velcro on a small pocket on the front of my armor, under my chin. It crackled and popped as the pouch slowly opened. I pulled out a tiny cleaning patch to wipe the dew from the lenses of my scope. *I might need to put that in a pouch with no Velcro.*

I sniffled back my runny nose and felt a sneeze coming on. I pressed hard on the infraorbital nerve under my nose to stop the sneeze. Sneezing would completely compromise the mission. *This cedar pollen is going to kill me.* I pulled a small pouch from the bottom of my ruck that I had labeled "Comfort." I kept it separate from my medical kit because it wasn't 100% necessary and sometimes it wasn't worth the extra weight or space to carry it. It contained a small bottle of sunblock, bug repellant, Excedrin, eye wash, Benadryl, and some other meds. I popped one of the Benadryl tabs to stop my allergic reaction and one of the Excedrin. The Excedrin was partly for the analgesic effect, but mostly for the caffeine it contained. I'd need it now to counteract the drowsiness of the Benadryl. I was preparing for a couple hours of sitting motionless before the sun came up.

"GO" TIME

When dawn broke, I got a boost of energy. I had been systematically flexing my legs, but they were both asleep. When it was time to get up, I would probably wobble like a newborn giraffe.

"Sierra One, OC-1," Lil Toe's voice echoed in my headset. I could hear the growling of the big MRAP armored vehicle in the background noise of his transmission.

"Go ahead for Sierra One," I answered.

"We are five mikes out. SITREP." He asked for a Situation Report.

"Someone is up. Lights were on before sunrise and the curtains moved on white-alpha-three thirty-two mikes ago. No PID. The car in the driveway has a tag that comes back to our guy." I let him know that I didn't have Positive Identification of *who* was in the trailer, but *someone* was, and it was probably our subject. The subject had made it known to his friends that he would rather shoot it out with the police than go back to jail. We took that sort of talk seriously.

Lil Toe asked Dangler his SITREP as well, then announced, "Last turn. Thirty seconds out."

I zoomed back the magnification on my scope so that I had a wide field of view and could see as much of the objective as possible. If I needed to take a closer look at something, I could just zoom back in. I checked that a round was chambered in my rifle and that the safety was on.

Now I could hear the crescendoing growl of the MRAP. The big armored beast smashed through the fence at the driveway, and the gate crunched under the MRAP's weight. Our trusty old white Suburban followed the MRAP to the front of the house. Next our two up-armored military-surplus HMMWV's (Highly Mobile Multi-Wheeled Vehicle, pronounced Humvee, a.k.a. Hummer) came ripping up behind them, engines roaring. They were loaded with a veritable cornucopia of tactical paraphernalia: breaching rams, bang poles, breaching shotguns, 40mm gas launchers, tow straps, armored shields, flash bangs, smoke grenades, and anything else they could fit in the trucks. The two Hummers split left and right.

There was no direct route to the back of the structure because there was a foundation from a long-since demolished building and a large tarp-laden shed blocking the way. In spite of the obstacles, the guys in the Humvees didn't waste any time getting to the back of the structure since their assignment was to keep the subject from escaping out the back of the trailer and into the wood line where Dangler was concealed.

I could see Big Toe in the front seat of one of the Humvees with Gobbler. They looked like they were having too much fun. The path to the left of the trailer was narrow, because a barbed wire fence separated this acreage from the one adjacent. In the remaining space, there was a 5,000-gallon 'cheapo' pool that was half-full of stagnant green water. The big tan Hummer down-shifted and accelerated. I could see the grins of anticipation in their faces as they smashed into, through, and over the pool, generating a mighty mist of putrid green spray. They slammed to a halt past the pool and parked their armored vehicle in a position to block any escape from the left side of the trailer.

In fairness, there really had been no safe alternative way to get into position. Sometimes the tactical gods just give you no choice but to do what any hyperactive, unsupervised, nine-year-old boy would do, if the nine-year-old had access to a bulletproof 7,000-pound truck: Smash the shit out of that pool.

"OC-1, this is Spider One. Green side is secure," Big Toe's voice came across the radio from his Hummer.

"Yeah…I can see that," Lil Toe didn't seem impressed. "Sierra One and Two, I need you to collapse in. We need more bodies up here."

I could see Lil Toe to my left, 96 yards in front of me. He was pushing the press-to-talk button on his armor and scanning the woodline to try to find me. Between Dangler and I, we could see the entire team, all four sides of the trailer, the Hummers on either side, and the MRAP and Suburban in front. The subject in the trailer knew for certain that there were four vehicles around his house, but he had no idea that snipers were watching from the woods, invisible and ready to engage any threat immediately.

"OC-1, Sierra One. I have an excellent overwatch hide. Sierra Two and I have you completely covered. I recommend we keep Sierras in place," I said.

"Negative, Sierra One. Move up here. Now," Lil Toe replied.

There were times to argue when you disagreed, but this was not one of them. I unloaded my bolt action rifle, collapsed the tripod, and packed everything into my ruck. I checked to make sure my compact M4 carbine was loaded and on safe and that the optic was on. I slung my ruck on my back, then my bolt gun, and cinched my helmet down. I secured my M4 tight across my chest. It was time to move, and I didn't want anything loose while I was sprinting.

I keyed the press-to-talk button on my chest. "OC-1, Sierra One. I'm Oscar Mike (on the move). Cover my movement from my hide to your position."

Lil Toe answered back quickly, "You're clear to move." There was no point of cover between me and the armored vehicles. There wasn't even a secondary route that could conceal my movement. I was counting on the subject having never seen me in my sniper's hide. At best, he would see a man in uniform come running out of a cedar tree and get behind an armored vehicle before he could even make the decision to shoot. I was usually one of the slowest distance runners and swimmers on the team, but I was the quickest on the sprints, even though we had to wear armor and a helmet and carry a rifle during the timed sprints. I loved sprinting in general, but sprinting with weight on my back was my jam.

Kneeling on my left knee, I keyed my mic. "Sierra One, moving now."

I trusted that my teammates had rifles pointed into every window on the trailer. If they saw a muzzle flash while I was running, they could at least suppress his fire. If you aim for the lower left corner of a window from which you see a muzzle blast, there's a good chance you're going to put an end to the shooter. Most people shoot right-handed, and they tend to stay behind the wall and shoot from the window sill. So, the lower left corner was where you could usually expect to return effective fire.

I tucked the butt of my carbine in under my right armpit and pointed the muzzle up. I shifted all my weight onto my left leg and launched out of the cedar tree. I locked my eyes onto the MRAP as I leaned in and accelerated. After sprinting the length of a football field with 70 pounds of gear on my back, I skidded in behind the MRAP, took a knee, and quickly stripped off all my sniper gear. It was time to change hats. Lil Toe needed me to be an Assistant Team Leader now.

LET ME CALL MY GIRLFRIEND FIRST

Our negotiator keyed up the PA from the safety of the front passenger seat of the MRAP, which was to my left. The Suburban was on my right, where teammates had taken up protective positions behind the engine block and wheels. I had moved to stand behind a large oak tree between the vehicles. I put the fingers and palm of my left hand on the tree, with my fingers facing toward the sky, and leaned my weight into it. I made a "V" with my thumb and index finger and rested the forend of my carbine in the "V." It wasn't ideal, but I had a steady shooting position that I could hold without too much effort, and I was protected by the trunk of the mighty oak.

The PA speaker on the MRAP screeched with feedback for a second before I heard the voice of our negotiator, Jose. He was a friend, fellow pilot, and fellow service academy graduate. He called the subject by name in his usual calm, steady voice.

"Thomas Knifechief, we have a warrant for your arrest. Come out now with your hands up." But there was no response.

Suddenly, the curtain moved on the window on the front door.

"You see that?" I asked Lil Toe, watching the curtain through my scope.

"Yeah," he answered.

I wasn't sure if the subject was just peeking – or if he was moving the curtain out of his way so that he had a clean shot at us.

After Jose informed the subject over the PA that we would bring him a phone, we formed a shield team and approached the front porch. We placed a throw phone by the door so the negotiator could have a two-way conversation with the subject. This was a very different *modus operandi* for us. Prior to having armored vehicles, we relied on speed and surprise to execute a high-risk warrant. Now that we had armor, the pace was slower and more deliberate. We no longer had to worry (as much) that we would get shot as we stood in someone's yard or on his porch. The pace felt too slow, but it was hard to argue that it wasn't better in general to just surround and call someone out versus kicking in the door right away.

It was certainly less exciting than the execution of warrants from my earlier years, but it was less likely to end in injury for us or the subject, and that was always the goal.

After we backed away from the door and got behind the trucks again, the negotiator called over the PA, "The phone is on the porch. You can get it now. I promise you no one is going to hurt you. You can take all the time you want. We won't come in to get you. If you understand, move the curtain on the front door again."

I resisted the urge to repeat the old joke where a sniper says to a negotiator, 'Tell the bad guy to come to the window and show us his gun and the hostage,' at which point the sniper has an obvious opportunity to make a canoe out of the bad man's head. There was a time to make off-color jokes, but this was not one of them.

Two sentences that the negotiator had used were unprecedented and worried me. The negotiator meant well to de-escalate the situation by saying, "No one will hurt you," and "No one will come in to get you," but it dissolved some of the subject's fear that the door would fly off the hinge at any second and operators would pour in to drag him out. We wanted to avoid a gunfight, but we needed to keep pressure on the subject. He needed to feel that he was always in jeopardy, and that we alone had the initiative. Those words from the negotiator might have de-escalated tensions, but they could backfire.

The subject and the negotiator continued to chat on cell phones while the negotiator sat in the air-conditioned armored vehicle. Sweat poured down my face and dripped on the inside of my sunglasses. I could feel my heartbeat in my neck. I was sure my heart rate was elevated because my body was straining to cool down. *I have to hydrate.* I drank the hot water from the bottle that was in my left cargo pocket.

The conversation continued on-and-off for over an hour. The local sheriff's office that had come out to secure the perimeter for us started packing up and leaving. Usually the FBI would show up, smash the door, flood the house with operators, and drag out a stunned suspect before he had a chance to mount any sort of fight. The current situation was not the show the sheriff's deputies had come to see.

"Hey, Lil Toe. We should gas him," I said.

"Not yet. The negotiators think that could unnecessarily start a gunfight. We don't need another gunfight. We have time on our side." Lil Toe said.

"We know he's the only person in the trailer," I said.

Lil Toe didn't answer.

Another hour went by. The guys were tough, but heat injuries are life-threatening. Big Toe was the senior medic, but he was in one of the Hummers on the perimeter. It fell on Dangler and me to look after the health of the guys. I left the safety of my big oak tree to make my rounds.

One at a time, I convinced my guys to drink whatever water they had – even though hot water is disgusting when you've been in 100-degree heat for hours. The lack of sleep combined with the heat was making some guys show signs of heat exhaustion. My worry was that most of them would power through all the symptoms until they just passed out. I didn't have any saline bags or IV fluids, and even if I did, finding a vein on an unconscious, dehydrated man was a situation I wanted to prevent before it occurred.

I had learned that lesson the hard way after an extended mission in the woods on another August day six years ago. We had set up observation and arrest teams while we waited for the subject of an extortion case to pick up a suitcase with millions in cash. He'd taken the bait, but he'd arrived two hours later than expected. In the meantime, my team had been in an abandoned shack and another team was camouflaged in the woods. It had been 110 degrees ambient and several guys had taped off their ankles and wrists to keep out swarms of ticks. It slowed down the ticks, but it also reduced the circulation of air they needed to cool off. I had insisted that guys drink their hot water that day, but some refused.

Once the subject had been put in custody and the cash had been recovered, we had made it back to our vehicles, but Wylie had had to be taken to the ER to get IV fluids. Unless it was a hostage rescue or a kidnapping matter, it wasn't worth dying from heat stroke. After that, when I told a teammate to drink water, I didn't take "no" for an answer.

"Hey, Lil Toe. We should gas him out," I said again in a deadpan voice.

"Noted. Stay on your gun. The negotiators are working. They said he's ready to come out in fifteen minutes."

"They said that two hours ago."

Lil Toe didn't respond.

I scanned the Hummers on either side of the building. We called the vehicles "Spiders" because it was our team's mascot. When Big Toe had designed the Oklahoma FBI SWAT patch, he asked a member of the Cherokee Nation how "SWAT" would translate into Cherokee. There isn't such a word, so they gave us the Cherokee syllabary spelling for "Secret Fighter," or better yet, "Quiet Warrior." That was perfect. Our culture was to be brave and bold, but humble and to put others first. On one occasion, we had been providing training to the Cherokee Marshals on man-tracking when a trainee had looked at my patch and said, "Secret Spiders?" One of the Cherokee symbols had a little squiggle on the bottom that he hadn't seen. Without that squiggle, it looked like the Cherokee word for "Spider," not "Warrior." So, we adopted the spider as our team mascot.

I keyed up my radio to talk to the Hummer drivers. "Spider One and Spider Two, this is OC-3."

"Go ahead for Spider One."

"Go ahead for Spider Two."

"Do you have a good angle for deploying gas?" I asked.

"One does."

"Two does."

"Copy, stand by," I said.

As long as the angle and range were good, the Hummers could fire 40mm gas launchers from the relative safety of their top turrets.

"Hey, Lil Toe. We should…"

Lil Toe's eyes darted over to me and he scowled. "Don't say it."

I paused for effect. "…gas him out."

"You're pissing me off, Money. Be patient."

His job was to make hard decisions and execute the mission. My job was to advise him and look after the health of the men. "It has been four hours." I said calmly.

"Jose said the guy agreed to come out after he called his girlfriend."

"He said that the last three times. He's screwing with Jose."

Lil Toe didn't answer and I went back out to make rounds and check on the guys. We had an ambulance on standby and I was worried one of my guys would be in it soon.

It was unprecedented to let negotiations carry on this long. But every thirty minutes or so the negotiator would say, "That's it. He's coming out." Followed by, "Hold on. He just called me. We need to wait." Six hours went by and everyone was exhausted. Actually, everyone who wasn't in an air-conditioned armored vehicle was on the verge of passing out.

During this entire time, two ASACs were about a half mile down the road sitting safely in their cars with the air conditioning running and listening to the events unfold on their radios. One of the ASACs was very flamboyant. He often wore cowboy boots with his business suit. When he sat down and crossed his legs, you could see that the tops of the cowboy boots were embroidered with a large red flaming eagle. He also had a custom holster for his pistol that had the FBI badge embroidered into it. He had never served on a SWAT team, but very much enjoyed being involved and alluding to others that he was "with SWAT," although he was only a managerial liaison.

Some of the guys were reflexively jerking their heads as they nodded off while they were looking through their rifle scopes. The subject called his girlfriend and told the negotiator that he was coming out. This time it was "for real." It was the ninth time.

BAD ROBOT

Lil Toe had had enough. He keyed his mic. "Negotiations are over. Prep for an assault." It was his call to make, and he had been more than generous with our time. It was 4:30 in the afternoon – seventeen hours since the team was activated. We usually worked very well with negotiation teams, but enough was enough. Sitting in an air-conditioned armored vehicle gave the negotiators a lessened sense of urgency. Their counterparts in the air-conditioned NOC (Negotiator Operations Center) trailer down the road had the same confidence to extend the negotiations as long as possible. From their perspective, a long negotiation was inconvenient but extremely safe, since there would be no assault on the trailer. But heat injuries from fatigue and dehydration are real injuries as well.

Lil Toe had us form a shield team with Vulcher on the shield and me on rifle. Wylie was behind us with the "throw-bot" and Lil Toe was behind him. As we approached the open front door, Wylie threw one of the robots and it fell short by about six feet. That was shocking because Wylie had a hell of an arm. I think dehydration was getting the best of all of us. Luckily, Lil Toe had a second robot. Wylie threw that one, and it also fell short.

I was astonished. Lil Toe seemed like he was coming unglued with frustration. Without telling anybody, he just walked around the shield team, picked up one of the robots and threw it through the front door. Then he walked back and got behind the shield team. It was brave, but definitely not safe.

We stayed behind Vulcher's shield on the front porch while Wylie cleared the front room with the robot. He got a good view of the first room with the remote control before the robot got stuck on trash. Bad guys rarely kept a clean house. Getting the robot stuck on trash and dirty clothes was the norm.

"Bang and clear," Lil Toe ordered. Wylie pulled a flashbang from his pouch and held it forward so I could see it. I gave him a nod without taking my eyes off of my optic. As the bang flew into the front room, I stepped swiftly forward and launched through the front door. *BA-DOOM!!* It detonated as my heel struck the ground inside the threshold.

Vulcher dropped his shield and moved behind me, gun up. With the front room clear, we methodically moved to the next room where we found the rapist. Since he had been stalling this long, was armed, and was known for violence, I was certain we would be running into a gun fight.

Surprisingly, he surrendered and had already put his gun down. Maybe when he saw the four of us, dirty, tired, and pissed off, he knew this was a fight we didn't want, but were prepared to end. The trailer reeked with the smell of rotten food and body odor. Piles of filthy clothes and putrid garbage swarmed with bugs.

"Covering," Wylie said as he pointed his rifle at the subject.

"Cuffing," I answered. I cuffed the bad guy behind his back, patted him down, and escorted him out to a waiting patrol car.

UNCONTROLLED SLUMBER

We began to collect our gear and get ready to go home. We were sharing what water we had to make sure that everyone had at least a little bit to drink. The day had reached 110 degrees with 80% humidity and we'd had no shade.

As we packed up our Suburban, another truck rolled up next to us. The ASAC with the cowboy boots jumped out and walked toward us, and said "Hey, can you guys grab me a bottle of water? I'm parched."

Lil Toe and I stared at him in disbelief but gave him one of our last bottles anyway.

As the MRAP delivered me to my truck where I had begun my stalk with Dangler eighteen hours prior, I looked like I had just gotten out of a swimming pool. There was no part of my body that wasn't soaked in sweat. I stripped off all my gear and my uniform top. I hated driving home in a wet, dirty shirt, so I put on a clean shirt that I always kept in my truck. I cranked up the A/C in the truck and sipped on a bottle of hot water from the center console.

I started to drive off but stopped and put the truck in park as I uncontrollably nodded off for a second. My body was now completely disobeying me. I slumped onto the steering wheel and let myself sleep for a few minutes as the engine ran and the cold air dried my uniform. My next stop was for a very cold, very caffeinated beverage. During the drive home, I kept the A/C high enough that it was uncomfortable and cranked up music as loud as I could stand. I needed to get home, but I was exhausted. I literally punched and pinched myself all the way home to stay awake. I had been up for about 35 hours and was dehydrated despite my best efforts.

Once I got to the house, I showered, climbed into bed at about 6 p.m., and slept until midnight. I woke up, wide awake and ready to go, so I went to the office and started catching up on reports from my cases. There are no "regular" hours in this job if you were on SWAT, and there is no paid overtime. It was just working cases and waiting for the next call-out. It was exhausting and brutal some days.

And I loved it.

****

The next Christmas, I sent the following poem to my teammates:

'Twas the night before Wewoka, when all through the Team,
Not an operator was stirring, not even Mean Gene.
The rifles were loaded by the 'Burban with care,
In hopes that the Lil Toe soon would be there.

The breachers were nestled all snug in their beds,

SEND ME

While visions of black Velcro danced in their heads.
And Dangler in his multicam, and Money in his cap,
Were the only two Sierras who weren't getting a nap.

The gentle droning overhead, while stumbling through dark,
Told us Voodoo was on station to watch o'er like a lark.
My commo went down; that was no surprise.
Surely Lil Toe was cursing and rolling his eyes.

Our IR beacons glimmered like the new-fallen snow,
While Big Al fired lasers at the objective below.
When, what to my wondering eyes should appear,
But two up-armored Humvees and a shit-ton of gear.

With an OD-clad driver, a lively old fellow,
I knew in a moment it must be the Big Toe.
More rapid than eagles his coursers they came,
And he whistled, and shouted, and called them by name!

"Now, K2! Now, Twilight! Now, Gobbler and Bodie!
On, Nestle! On, South Beach! On, Lobos and Mackey!
Load up the 40 and aim at the wall!
Now gas away! Gas away! Gas away all!"

Our subject, he slumbered, and ignored the PA.
Instructions were given to wait there all day.
Then up to the house-top the coursers they flew,
With the sleigh full of negotiators, and an ASAC or two.

As I flipped off my safety, and was aiming for certain,
I heard him announce, "Please move the curtain."
Off my scope I did come, and let out a yawn.
"Let's get going," I said, "I've been here since dawn!"

His first promise was: "No one will hurt you."
"You can wait there all day; we'd never come get you.
Now move the curtain once more, and please take your time,
and do call your girlfriend, we'll lend you the dime."

Heads were then nodding and eyes were quite heavy,
My rifle felt as hefty as a '66 Chevy!

Eventually the words were finally said,
"He's coming out,"…but he didn't instead.

One more call he'd make, or maybe t'was nine.
The negotiator said that was perfectly fine.
Until finally patience and sleep-loss wore thin,
And word was given that we would go in.

Vulcher with the shield, and I at his side,
Wylie, a robot, he sent as our guide.
Yet gravity seemed extra heavy that morning,
As the robot fell short without any warning.

A second brave robot was ready and resolute.
Wylie launched it as well, but it followed suit.
Cuss words abounded and gnashing of teeth,
And the smoke, it encircled Lil Toe's head like a wreath.

Suddenly he moved without making a sound…
Lil Toe marched up and set the robot aground.
The trailer was nasty and belonged in a landfill,
And all operators were ready to link up and exfil.

The ASAC approached, his stride full of bolster,
His words were flamboyant, as was his holster.
No fear had this man, his career without stain.
If you had any doubts, he'd tell you again.

A wink of his eye and a twist of his head,
Soon gave me to know I had something to dread.
He spoke not a word, but went straight to his work.
His boots bore flaming eagles; he turned with a jerk.

He sprang from his sleigh, to the team gave a whistle,
And called out in sorrow, "I'm dry like a thistle."
He grabbed him some water, 'ere he drove out of sight,
"Tear gas to all, and to all a good night!"

Chapter 6
Rocket Bomber
Mission #67
March 2014

The name "Lil Toe" flickered on my phone while it buzzed in the cup holder of my Suburban. A call from Lil Toe usually meant adventure was ahead. I grinned to myself a little before I pushed the green button.

"What's up, my dude?"

"Hey, J-Money. We've got a WARNO. It's an interesting one. We'll need to deploy snipers. I need you to come down here for mission planning."

I pulled off the road and cut a U-turn. "You bet, brother. I can be there in about 90 minutes." I got friendly teasing from my fellow Agents for having my seven-passenger truck so loaded up with gear that there was only room for one passenger, but this was yet another moment when I was happy to already have everything I needed.

I had been on my way to meet with Eva, a CIA Case Officer, who was my counterpart on a classified joint operation that we had been working on for over a year. I had been reassigned from Criminal to Counterintelligence (CI) investigations after 9/11, but I had gone kicking and screaming. Even though I missed working fugitives and bank robberies, I soon learned that CI work had challenges and rewards of its own.

By law, the FBI is the lead agency for exposing, preventing, and investigating foreign intelligence and espionage in the U.S. The CIA, despite its tremendous capabilities, is not. I hadn't always seen eye-to-eye with my counterparts in the CIA (a.k.a. "the Agency"), but in this case, I had really clicked with one of their Case Officers. In fact, our joint double-agent operation had been so successful thus far that our intelligence products were being briefed directly to the President of the United States. There was often tension between the Bureau and the Agency. After all, we were spy-hunters, and they were spies and spy-makers. They often thought of us as knuckle-draggers who only saw in black and white, and we often thought of them as manipulators that lived in the "gray." Maybe both stereotypes were based a little bit in truth, but a lot in hyperbole.

The CIA was critical to this operation, and neither Eva nor I made unilateral decisions. We worked as equals, and as a result we made huge strides in our intelligence work.

Working CI meant that I traveled a lot and worked a lot of weekends and evenings, but it also meant that I didn't have unscheduled call-outs for reactive investigations like bank robberies. My work was *proactive*, so I could plan my work schedule in advance, with the exception of unexpected SWAT call-outs. This was one of those times.

I dialed one of the numerous cell phone numbers that Eva had given me.

"Eva. Hey, it's Jeremy. How's it going?"

"Jeremy! Hey, I was just about to call you. I'm so sorry. Something came up and I can't make it today. I'm not even in Oklahoma right now," she said.

"Really? That's actually perfect. I was calling to reschedule, myself." We both laughed for a second. "Let's catch up early next week."

"That sounds great. Talk to you soon!"

That was easy.

In addition to being in sync with Eva, I was lucky to have a great supervisor, Trey, who let me do my job without breathing down my neck. As long as I was productive, he left me alone. As a result of that mutual trust and respect, I produced more national Intelligence Information Reports (IIRs) than any other Agent in the division. I had the freedom to get my job done on my own schedule and still be available for SWAT ops on a moment's notice.

OPS BRIEF

"Listen up." Lil Toe scowled and scanned the team room to get eye contact while mentally taking attendance. It smelled like gun oil, farts, and men's deodorant. Guys were leaning back in chairs behind rectangular tables or sitting inside their designated equipment lockers. Some guys were wearing shorts and t-shirts and others were in suits and ties. We had all stopped what we were doing to respond to the WARNO. Some of the guys had just gotten out of court, some had been at the gym, some on surveillance.

Lil Toe cleared his throat loudly to quiet the small talk as Radio, our team's Electronics Technician, dimmed the lights. The room fell silent when the projector lit up the front wall with an image of a house surrounded by thick forest on the edge of a lake.

"We have a federal search warrant for a subject who lives in a rural homestead. He has an extensive criminal history that includes charges for drugs and capital murder. Thirteen months ago, a bomb was sent to a sheriff in Arizona. Phoenix Division has identified the bomber as this guy..."

The face of a middle-aged man was projected on the screen.

"...a loner who lives in the woods in rural Oklahoma. This won't be a routine search warrant. He's well-armed, he has three Rottweilers and potentially has more bombs. There's no easy way to approach the objective without being noticed."

He was right about that. We couldn't drive the MRAP up to the front door on this one. That beast was so loud, you'd hear it coming from miles away, and the bomber would have time to squirt into the woods or set up an ambush.

The MRAP was a Mine-Resistant Ambush-Protected armored vehicle that was designed by Oshkosh to fulfill a Marine Corps military contract. However, they rejected these vehicles, which were then put up for grabs as military surplus. The FBI obtained one MRAP for each of the fifty-six FBI SWAT teams. They were high-maintenance machines, but they were free. Before we got them, the only armor we had to protect us from pot-shots was the engine block of whatever vehicle we used for arrests. The MRAP was a 14-ton behemoth that was made for taking a beating, but stealthy it was not.

Lil Toe continued, "The bomber lives on a spacious, isolated, rural property that is surrounded by forest. The south edge of the property ends abruptly with cliffs that drop 1,000 feet down to a 3,000-

acre lake. Overhead imagery shows that he has a shooting deck on the second floor of the house.

"Team leaders, split up and spin up your game plans. We will meet back here in an hour."

The Ops Brief was essentially the STL giving the team leaders the problem, and we brainstormed and came back to brief the solutions. As was our routine, the two assault team leaders and the Sniper Team Leader would meet with their teams separately and draw up the mission plans. The STL would give Blue and Gold Team Leaders our respective responsibilities. For example, Gold might be responsible for setting isolation around a structure while Blue was responsible for entering and clearing the same structure. Or Blue might be responsible to plan a hasty emergency hostage rescue plan, while Gold formulated a deliberate rescue plan that involved collecting blueprints, setting up a mock structure, and rehearsing the plan. The two teams were always set up to support each other, and those roles typically rotated every mission.

NO GROWNUPS ALLOWED

No supervisors were allowed in the team room during mission planning. Most supervisors didn't even have the code to the team room, and they hated that they had to knock to get in. SWAT operators are Special Agents in the paygrades known as GS-11 to GS-13. Once you chose to become a Supervisory Special Agent (SSA) at GS-14, you no longer did "street Agent" work. No more conducting investigations, flying airplanes, providing firearms training, conducting arrests and searches, testifying in court, collecting evidence, or conducting crisis negotiations. We used the term "street Agent" typically to mean that you solved crimes and put your main focus on being an investigator. There were, however, Special Agents that were not "street agents," even in the GS-11-to-GS-13 ranks. Those were full-time recruiters, polygraph operators, pilots, surveillance team members, outreach coordinators, and such.

Once you were a GS-14 or higher, you were relegated to a desk job providing oversight to squads and programs and managing budgets, personnel, and inspections. Since everyone on the team was essentially equal, no one in the room could pull rank and override the decision-making process. This was a fundamental key to the success of the FBI SWAT program.

After the whole team reconvened and team leaders had recommended their COAs (Courses of Action), the various plans were unified and integrated into an overall plan. Everyone gave input, from the newest operator to the most senior. Several options were weighed and measured until the team agreed on the safest plan with the highest likelihood of success. It took a tremendous amount of effort to put together a plan that made sure no one was injured or killed – not even the suspect.

It was a casual environment that usually involved a lot of dry-erase ink spread across several white boards while surveillance images and blueprints were overlaid from a projector. Everyone got to point out concerns and ideas until we reached a consensus. Once we were done, there was buy-in from the entire team. On rare occasions, if there was a disagreement, the STL would cast the deciding vote. The briefing, mission planning, gear prep, and rehearsals could take several hours, but the attention to detail for every op is what kept us safe. There were no "routine" operations.

MAIL BOMB

Our subject for this op had crafted at least one bomb and shipped it across the U.S. Luckily, a postal employee thought the amount of hand-postage on the package was suspicious and took it back to the post office. FBI Special Agent bomb technicians and US Postal Inspectors examined and disrupted the device before it could get to the intended victim, a county sheriff in Arizona. The FBI Laboratory had painstakingly examined every detail of the disrupted bomb and forwarded us the specifics.

The bomb was crude – made from off-the-shelf smokeless powder with a model rocket igniter to initiate the explosion. The igniter needed a D-cell battery to fire, so the bomber fashioned a simple circuit, stapled in place, that would close when the box was opened. Once the circuit to the ignitor closed, the main charge would detonate. He used pre-made boxes like those provided by FedEx or the US Post Office that closed with adhesive and were opened with a pull strip. Pulling the pull-strip detonated the bomb. Based on all the evidence, we obtained a federal search warrant to search for further evidence in the subject's residence. This would not be a "knock and talk" done by street Agents. It would be safer to send our team with armored vehicles and embedded bomb-technicians.

SNIPER BRIEF

I was Gold Team Leader and Sniper Team Leader, but I let my Assistant Team Leader run Gold for this op, and I was focused on leading the sniper team. As Sniper Team Leader, I briefed last. I walked to the front of the room to brief the sniper mission COA. I had been on the team for almost fourteen years by now and was a senior operator, but I never enjoyed briefing an operations plan. I made a mental note to stand still and speak clearly. I was the guy always making everyone laugh, so it was important now to show the team that I was "switched on."

"My recommended Sniper COA requires sending in a pair of two-man sniper teams, under cover of night and well in advance of the assault teams, to get eyes on the objective and collect intelligence. There's no good route to hike in from the north without being observed by several other rural residences. Our plan is to meet with the State Patrol on the south side of the lake and use one of their patrol boats to transport the sniper teams, as well as a three-man QRF (Quick Reaction Force), to an insertion point at the cliffs south of the objective. From the cliffs, we will travel 800 meters northeast through cedar forest and briar to establish sniper hides.

"I'll take the white-red angle and Dangler will take the black-green vantage, so we will have eyes on the whole structure long before you boys show up. We'll collect intel and provide lethal overwatch for you if the operation goes kinetic. Since the subject is armed and dangerous and has attack dogs, the QRF will hike in with the sniper team and be available for emergencies."

No one said a word. I looked at Lil Toe and he was expressionless.

I broke the silence. "Any thoughts or concerns?"

"Nope," Lil Toe answered. The guys were nodding and they were starting to fidget. These planning meetings sometimes took hours and none of us liked to sit still.

"Don't run off," Lil Toe said. "I'll have assignments for a bunch of you. We need to prep vehicles, coordinate with the troopers, and get aerial imagery."

AERIAL RECONNAISSANCE

"Clear prop!" Big Al called out the window to warn unseen bystanders before he hit the starter button on the aircraft panel. The big motor rattled and banged and the propeller spun to life. I already had my headset on and was buckled into the right seat. It smelled like 100-octane aviation fuel and old upholstery. I loved that smell. Al put the coordinates of the objective into the plane's GPS while I finished the engine-start checklist. It was just a few days before the operation, and we were going to confirm our intelligence about the bomber's cabin and its surroundings.

"You want to do the take off?" Al asked, his voice crackling into my headset.

"Don't threaten me with a good time!" I said with a grin.

I wiggled the yoke so Al would know I had the controls.

"My plane," I said, per protocol.

"Your plane," he replied, as he looked out the left window.

I dialed a four-digit "squawk" code into our transponder that let the control tower know that we were a "Voodoo" plane. There was no need to say it over the radio and tell every aircraft in the area who we were.

I keyed the mic button on the yoke and said, "Tower, Foxtrot Delta Three Zero Two, ready for take-off. Holding short runway one-seven, full length."

The tower replied, "Three Zero Two, cleared for takeoff, one-seven."

I eased the throttle forward and stepped on the left rudder pedal to bring the nose around and onto the active runway. As I lined up the main wheel with the striped centerline on the runway, I steadily pushed the throttle forward to the firewall and the engine roared. I loved the sensation of being pushed back into the seat. I glanced at the gauges as we accelerated. *Engine's RPMs are in the green. Exhaust temps in the green. Fuel flow nominal. Airspeed's alive. Need a little more right rudder. There's rotation speed. Light back pressure on the yoke.*

The grumbling of the tires over the concrete suddenly went silent, and the nose wheel shimmy stopped as the plane broke free from the ground. I eased some forward pressure back into the yoke and let her accelerate to the best climb speed before teasing her nose up again. The evening sunset was even more beautiful as we climbed.

"Nice work." said Big Al, approvingly. "Take us up to surveillance altitude."

I spun the trim wheel to hold us at a steady climb out.

"Thanks, brother. 'Maintain thine altitude lest the ground rise up and smite thee.' Right?" I said.

We took turns flying the plane until we got near the objective: the bomber's residence. Al throttled back the engine, tweaked the propeller setting, and lowered ten degrees of flaps. At our altitude and engine settings, we would be effectively silent to anyone on the ground. As Al put us in a lazy circle above the target area, I activated the FLIR (Forward Looking Infra-Red) pod to get visual-spectrum images and thermal images. As suspected, there was only one road in from the north and the lake to the south. It would not be easy to sneak up on this location. The deck on the second floor looked like the shooting platform that witnesses had told us about. Inserting snipers from the south before the assaulters drove in from the north still seemed like the best plan.

Reconnaissance completed. The wheels lightly chirped as Al set the plane down at the small airport. Here the plane would be fueled and stored in a hangar, paid for by an undercover entity. For all anyone knew, we were just two guys in shorts and t-shirts out for a joyride. As the hangar doors groaned shut, I finished downloading the surveillance images and sent them to Lil Toe with a message: "Confirming Sierras will need maritime insertion from the south. IMINT (Image Intelligence) inbound."

K2

My Assistant Team Leader (ATL) for Gold team was K2. He was a fellow U.S. Air Force Academy grad, and although our attendance overlapped for two years, we didn't meet there. Instead, we met in Norman, Oklahoma. I was assigned to the Norman FBI Resident Agency, and he was a commercial instructor pilot at the airport where I kept my plane. We both loved aviation and quoting funny movie lines, and we hit it off right away. I encouraged him to apply for the FBI and a year later, he got accepted. After Quantico, he got assigned to an FBI office in Texas where he tried out and was selected for SWAT. You can imagine my surprise when he called me eight years later to tell me that he was transferring to Oklahoma and wanted to transfer to my team. I liked to claim that I was responsible for encouraging a guy as sharp as him to join the ranks of the FBI.

IT TAKES A TEAM

K2 had briefed the operation to state and local police since he was most familiar with the area and the resources therein. We couldn't do our job without state and local police, sheriffs' deputies, firefighters, helicopter services, ambulance services, and hospitals. The operation was near his FBI satellite office, the Tulsa Resident Agency, and he already knew all the locals. At his request, the State Patrol had agreed to insert our sniper team into position with one of their patrol boats. K2 also coordinated prisoner transport with the local sheriff's office.

The closest medivac service was going to pre-position a chopper near our objective several hours prior to the operation so that we could get swift medical airlift for any injured operator. The local fire department was ready to stage an ambulance crew near the TOC (Tactical Operations Center), and K2 talked to the charge nurse at the closest Level One Trauma Center to coordinate discreet arrival of any injured operator.

As we did on some occasions, we left our wallets in the trucks and wrote our numerical identifiers (in my case, OC-2) on our left arms with permanent markers. This allowed us to be correctly identified at the pre-designated hospital without providing our true names. So, if as lead medic, I had transported wounded teammates to a hospital, the staff could tell me, "OC-14 is in intensive care from a gunshot wound and OC-5 has a head injury." I would know exactly to whom they were referring, and I could notify my teammates' families. But the staff wouldn't know our real names or identities until we had time to vet the medical staff who were providing care. I had a list of everyone's blood type, meds, and allergies in a sealed envelope in my trauma bag that I could provide to the medical staff. I even had each operator's preference for who and how we should notify their spouse about their injuries or death.

I always felt so grateful to get the local support that we needed. As a team, we chipped in cash to buy extra patches and unit coins to hand out to those who worked with us. They liked the swag, but I knew from experience that the local first-responder community would have supported us, no matter what.

There are over 400 small Resident Agencies across the U.S. in addition to the fifty-six large Field Offices. Those of us that worked out of Resident Agencies were known as Resident Agents, and we had a very different work experience. There were typically no supervisors, no support staff, and minimal equipment. In some cases, we cleaned our own toilets and vacuumed our own floors. It was a privilege to work in an RA. You could set your own hours as long as you logged at least fifty hours a week and put criminals in jail. Most of us hung out and worked cases with our local police. The RA office was essentially just the place to write reports. The RA supervisor was mostly just a guy we talked to on the phone and who signed those reports.

The reason for the FBI's effort to maintain numerous RAs across the US is that Resident Agents live in the community with the people they serve and protect. A lot of the local cops, nurses, and firefighters were not just colleagues at a time like this; they were our friends. Our kids went to school together, we went to church together, we would run into each other at restaurants. My twenty-three years of experience as a Resident Agent and SWAT operator was the antithesis of what Hollywood portrays. We weren't the detached men-in-black that commandeered a crime scene. We were fellow cops, medics, dads, and members of the community. The Resident Agency is a small outpost, but it's the heart and soul of an effective FBI.

MOTLEY CREW

One of the things I loved about the FBI SWAT team was the diversity of skill. As I sat in my locker in the corner, I thought about some of my teammates' previous jobs: Federal Air Marshal, Deputy U.S. Marshal, 75th Ranger Regiment, businessman, neuroscientist, Air Force navigator, pilot, Border Patrol agent, CIA "accountant," police SWAT, detective, US Marine, cyber expert, Navy SEAL, NSA, Navy bomb tech. We were a diverse and motley bunch. It sounds arrogant, but this was a pinnacle job. These guys had all left successful careers in other agencies to come to the FBI and try out for SWAT.

As Special Agents, we specialized in the most niche, singular career paths one could imagine: Weapons of Mass Destruction, Counterintelligence, Counterterrorism, Cyber Crimes, Transnational Organized Crime, White-Collar Crimes, or Civil Rights. As we sat in this room, we took on our additional tactical roles as team leaders, medics, breachers, snipers, bomb-technicians, and assaulters – or a combination thereof. If any group of people knew how to wear multiple hats and still be high-performers in every role, this was it.

As different as we all were, we all wore the same 14-karat gold-plated badges with solid copper-alloy cores. We all carried the same title, "Special Agent," as well. That term was first used in 1870 when U.S. Attorney General (AG) George Williams appointed a "special agent" in the U.S. Department of Justice to conduct "special investigations" for the AG. In 1908, AG Charles Bonaparte reorganized the Bureau of Investigations investigators into a "special agent force" that became the FBI that same year. In the 1930's, it was decided that all investigative agents – agents and accountants alike – were to be called Special Agents and the convention of capitalizing "Special Agent" became uniform in the FBI.

I'd like to think that those first Special Agents would be proud of the work we were doing now.

GEAR PREP

It was late in the evening the day before the op. My regular casework had required me to get some last-minute reports done at my second-floor office in a strip mall. I was going to testify at a Federal Grand Jury the following week for a multi-million-dollar embezzlement indictment. I had spent over a year on this case, and I needed to make sure that all my reports were perfect. There just wasn't time for everything, so I often gave up sleep to try to get everything done.

I've got to pack for this mission soon...then get some sleep.

I was tiptoeing around the house so I wouldn't wake up my daughter or wife. The spare bedroom closet door squeaked and rattled in its usual fashion. *I need to fix that door.* My uniforms had been banished to this closet so my wife could have at least a little room in our bedroom closet. I had an entire row of business suits and ties for court appearances that hogged up most racks. If I hadn't kept my uniforms in the spare closet, she wouldn't have had room for a single t-shirt.

I still had old woodland-camouflage BDU's (Battle Dress Uniforms) and flight suits in the spare bedroom closet, but we didn't wear those anymore. *I really should put those in the attic,* I thought. But I loved the memories each old uniform carried, and I liked seeing them in the back of the closet like retired old warriors.

I grabbed a Crye combat uniform in Multicam camouflage pattern. My teammates would be wearing our traditional olive drab green, but I opted to wear our alternate uniform because of its effective camouflage pattern. Since there was a fresh layer of snow on the ground, I grabbed thermal underwear and camouflaged Gore-Tex pants and jacket. Based on the forecast, most of the snow would probably melt and I would be spending a lot of time in the prone position, either in the snow or on saturated mud. Although the cold would suck the warmth out of me eventually, at least the thermals would be a barrier between me and the soggy ground. I had learned during SERE (Survival, Evasion, Resistance, and Escape training) that if you lay on the ground long enough, you eventually become the same temperature as the ground. And you die.

I dug through my sniper ready-bag and pulled out my face paint. I rarely painted my face, but this op seemed like one in which camouflage would be critical. In the morning, I would put the darkest colors (black and brown) on the high point of my face, like my nose and brow, to distort the pattern of a human countenance. I'd do it by the book, too: paint on my eyelids, nostrils, and in my ear canal.

Remember: You're going to look like a Goth punk wearing eyeliner and eye shadow until at least three showers. I made a mental note to remember to clean my eyelids after the op. I usually forgot on account of the fact that a person can't see his own eyelids in the mirror when his eyes are open.

The night before, I had touched up spray-paint camouflage on my body armor, leather combat boots, battle belt, gloves, holster, rifle, ruck, and everything that was exposed to the ambient air. I might look like an oddball when I stepped off the truck at the end of the day, but during the op, I would be invisible.

My ghillie suit stayed in my truck – as usual. The ghillie suit is an overgarment that is typically covered in brown, green, and tan jute and then painstakingly decorated with local vegetation to further protect the sniper from his enemy's view. Ghillie suits are popular in video games and movies, but in real life, stalking through a forest in a ghillie is like wearing a camouflage wedding dress made of Velcro – in a forest full of more Velcro. The term "ghillie" is said to have come from the Scots Gaelic word which means "outdoor servant." Scottish gamekeepers are thought to be the first to use them. And they can keep them. They are a pain in the butt.

I hit the garage door opener and walked out to my truck. I stood in the driveway in socks and Crocs and an old fuzzy, military camouflage coat. The freezing air was actually refreshing. I punched in a code on the vault keypad and pulled a big metal vault drawer out of the back of my Suburban.

Inside, I had a veritable arsenal to choose from: I wouldn't be using my compact Close Quarters Battle Rifle, my SR-25, my intermediate-range scoped AR-15, breaching shotgun, or my 10mm MP5 submachine gun. Instead, I did a "pre-flight check" on my primary sniper weapon, an HS-Precision .308 Winchester bolt-action rifle. I grabbed the torque wrench out of my bag and checked the torque on the scope and the action bedding screws. I checked the barrel fit and scope position and made sure the stock was in the position I preferred, and I marked it with red Sharpie.

I slid the big beast into my Eberlestock rucksack. It was a good choice for holding my rifle on my back since I would be climbing and maneuvering. *I wish I had had this pack at Glass Mountain.* The rifle weighed seventeen pounds and bore a twenty-four-inch heavy barrel with flutes that reduced weight and increased barrel cooling. It was a big pig to carry around, and a ruck to bear the weight was just the ticket. I pulled out four of my magazines and double-checked that three were loaded with 168-grain Federal Gold Medal Match (FGMM) and the fourth was loaded with Federal 165-grain Tactical Bonded rounds.

The bonded round was pretty impressive. It could defeat intermediate barriers like glass and steel without breaking apart and then expand into a .50-caliber slug once it hit soft tissue. I had taken the measure of several Oklahoma white-tail with it, and it had the authority of Thor's Hammer. The FGMM, however, was more accurate while still providing more than adequate terminal ballistics.

From the side pouch on my ruck, I pulled out my PVS-22 night-vision optic, put fresh batteries in it, and made sure it worked. I put on my helmet, checked the tiny white, red, and infra-red (IR) lights attached to it. Then I flipped down my night vision goggles. I checked my IR strobe as well as the IR flood light mounted to the side of my rifle. I had rigged a pressure pad from the IR light to a strap of Velcro forward of my rifle's safety, so I could illuminate a target without affecting my shooting position. My rifle had been issued to me with its factory black barrel, stock, scope, and sling, but I had long since changed that. A few layers of green and tan rattle-can spray paint muted the colors. There are few true black colors in nature, and an all-black rifle stuck out like a sore thumb.

The last item to inspect was my downlink. I popped open a water-tight Pelican box and slid the brick-sized mini-TV/receiver out of the foam cut out. It was a $10,000 item that let me see the exact same birds-eye-view that Voodoo saw on its thermal scanner in real time. Its size made it awkward to strap to my left wrist, but it was an evolutionary leap in situational awareness for snipers.

I dug the cigarette lighter charger out of my Pelican box, connected it to the downlink, and plugged it into one of my eight cigarette lighter ports. I didn't have any USB ports, but Radio had installed extra cigarette lighters and rigged three of them to be "hot," so that it would charge while my truck was turned off. I always had my GPS, flashlight, handheld thermal scope, and my two handheld Motorola radios charging. The radios cost $7,500 each. And somehow they trusted me with two of them. I would not put them in the pouches on the back of my armor until the last minute so that they would have maximum charge. Most operators carried one radio in the front on their nondominant side, so they could change the volume and frequencies. I had two radios: a short-range radio to communicate with my teammates, and a long-range radio to talk to aircraft and the TOC. I had them both on my back so my body didn't smother the signal when I was in the prone position or crawling. Having them on my back also freed up space on the front of my armor for flashbangs, rifle magazines, a tourniquet, and a small "get-the-hell-off-me" dagger.

The cold night air was soaking into my bones now. I thought about how fortunate I was to live in the countryside where I could do my gear checks in private. Otherwise, I'm pretty sure someone from a neighborhood HOA would have called 911 already.

I went back into the house and pulled out my secondary weapon: my trusty Glock 21. It was a big, fat .45-caliber pistol that held fourteen rounds. It had a light attached to the rail and was nestled in a Safariland ALS holster. I had fired over 40,000 rounds through this particular pistol. In its many years of service, my .45 had been sent back to the FBI weapons vault in Quantico five times for preventative maintenance, but it rarely jammed.

On separate occasions, I had managed to wear out the frame, the slide, and all of the internal parts. I kept a close eye on it when I cleaned it, and I had identified cracks in the breech face and frame before they got bad enough to cause malfunctions. At this point, it was an entirely new gun built around the original factory barrel. The octagonal rifling was still holding up and the weapon would shoot two-inch groups at twenty-five yards with Remington 230-grain Golden Sabre hollow-points. I had had the opportunity to harvest an eight-point buck with this cartridge one beautiful Thanksgiving Day and it performed perfectly.

Lastly, I checked all of my other equipment: a thermal scope, a spotting scope, bipod, compass, GPS, hydration bladder, "blow-out" first aid kit, and most importantly, "pogey bait" (a.k.a., snacks). I pulled the carbine magazines and flash-bangs off of my armor. I didn't need the extra weight. I had previously weighed each item of gear on an old postal scale and considered carefully its worth-to-weight ratio. The infil to the objective would only be 800 meters. Fully laden and fighting through briars and yet maintaining stealth, a healthy pace would be about 1,600 meters per hour. To be safe, I'd plan for an hour for the hike in. I had to consider how quickly I could cover that distance with my chosen load. Less weight meant I'd be faster and have better endurance.

I checked my watch. *10:30pm.*

With everything set for the morning, I grabbed my car keys and drove across the state to a hotel to get a few hours of sleep.

RALLY POINT

The wind was bitter, but it was keeping me alert. My 3 a.m. coffee was slowly starting to take effect. I was glad I had put on the extra layers of clothing, because the cold was going to catch up with me later. From my parking spot it was a short walk to the patrol boat. I shook the hand of a burly trooper, and we were making small talk when headlights lit up the dock. My teammates got out of their cars and walked over to us.

My spotter was Febreze. He had been a neuroscientist before he joined the Bureau. Despite being a big nerd who was fascinated with the human brain, he was built like an Olympic track-and-field athlete. He was a lot of fun to be around, a devoted family man, and always quick to laugh. He got his call sign when he roomed with Vulcher at SWAT Basic Training. Febreze only brought two uniforms for the three weeks of training. When Vulcher questioned his judgement, and expressed his concern about limited laundry options, and the impending room stench, Febreze pulled a giant bottle of air-freshener out of his pack and said, "Don't worry. I brought this!"

The other sniper was Dangler, and K2 was Dangler's spotter. Twilight, Wylie, and Bodie were the Quick Reaction Force. Twilight and I walked over to my truck, and I pulled my Heckler and Koch MP5SD out of my weapons drawer. It was a compact submachine gun with an integral suppressor. That means it can shoot a whole lot of very quiet bullets, really fast. Since there might be angry dogs to deal with, and we didn't want to give away our position, I gave Twilight my MP5SD. Shooting a dog that is defending his property would be a last resort, but if we were forced to do so, we wouldn't want to compromise the mission by not doing it – or being loud about it.

We walked back over to the group on the dock. Dangler and I went over last-minute medical emergency plans. We were both nationally registered EMTs in addition to being snipers, so we had contingency plans to switch roles and provide medical care if things went south. We hadn't been outside for long, and the chill was already starting to seep into my bones. *Time to go to work.*

I imagined the assaulters at that moment, scrolling on their phones in their heated armored vehicles for the long and not-so-stealthy drive in. Actually, they might not even be in their trucks yet. They were probably still drinking coffee and getting kitted up. The warrant wasn't for another three hours.

MARITIME INSERTION

The dual outboard motors gurgled in the murky water as my seven-man infil team loaded onto the boat. As soon as we were in our seats, the boat pulled away from the dock. The driver turned the bow into the darkness and moved the throttles all the way forward. My head pushed back into the headrest because of the acceleration. The rush from the speed gave way to a little concern when I saw the driver switch off his depth gauge and switch his screen to GPS. His gaze was fully downward as we ripped across the lake in the middle of the night. I guess there was nothing for him to see out the front window into the predawn darkness.

The engines roared like a pair of chainsaws and each small whitecap launched the boat into the air. Luckily, we were far enough from our destination that the subject wouldn't hear us. I hoped. I had not envisioned that we would be this fast and this loud. I had assumed we would take a slow, casual boat ride – preferably a *quiet* one.

I flipped down my NODs and saw a spray of light green water over a dark green lake with a black sky above. *Man, I bet that water is cold. We better not crash this thing.*

I hated the annual water survival test, but now I was glad for the training. Every year we had to tread water in full uniform for twenty minutes and swim 400 yards in full uniform with boots. We had to jump from the top of an Olympic high dive, activate our Mustang inflation devices, then climb back up to the high dive on a rope ladder. After that, we put on a helmet, body armor, and a training rifle and jumped in the deep end. You had to doff your gear while trying not to sink to the bottom. You couldn't just tear it all off, you had to maintain your composure and hand each item to the operator treading water next to you.

I had watched one teammate start by pulling his rifle sling over his head in an effort to doff it first. The sling had slid up his neck, under the back of his helmet, and pushed his helmet forward over his eyes. He got a mouthful of water and panicked. I never forgot that. *If we go for an unexpected swim, drop your ruck, pop off the helmet, inflate the Mustang. No need to doff your armor.* I was one of the team's worst swimmers. Only Mackey hated it more than me. For the most recent water test, he had brought a purple foam pool noodle and insisted he keep it under his arms the entire test. Everyone thought it was so funny that they let it slide. Except me. I was mad I hadn't thought of it first.

I flipped my NODs back up and closed my eyes while I visualized the navigation points that Twilight had mapped out and named for simplicity and brevity:

Camaro - Primary Boat Insertion Point.
Catalina - Alternate Boat Insertion Point.
Corvette - Halfway Point.
Trans Am - QRF ORP (Objective Rally Point).
Firebird - Objective.

Starting the infil at Catalina could have made for a shorter hike, but the topographical map indicated the cliff was significantly steeper than at Camaro. Twilight knew better than any of us how to read contour lines on a topographic map, and I agreed with his decision in selecting Camaro as primary.

My weight shifted forward as the engines cut back to idle, and the boat slowed to a silent crawl. I flipped down my NODs and walked out to the bow of the boat with Wylie. I could see a near-vertical 900-foot cliff in front of us.

"Man, that looks steep, doesn't it?" I whispered to Wiley.

He nodded. "Absolutely."

I reached for my GPS to confirm our position, but Twilight was two steps ahead of me.

"This is Catalina," Twilight said with mild disappointment as he looked at the boat driver. "We insert at Camaro."

"Yeah man, but this will be a shorter hike for you guys," the driver answered.

Wylie stared at the driver. I could tell he was trying to contain himself. "Look at that cliff," he said sharply. "We don't have technical climbing equipment with us. We need to go to Camaro."

The driver flipped a switch and suddenly blinding floodlights from the boat lit up the cliff like it was midday. My night vision goggles automatically dimmed to compensate for the floodlights. I hoped we were far enough away from the objective that the boat's lights hadn't given our position away. I hadn't considered that the boat driver didn't have night vision and had been looking into the void of the night while pulling up to the cliff. He meant well.

"Oh, shit, that's steep," he quipped. "Y'all want to go to the other place?"

We nodded silently in unison and shuffled back to our seats as he shut off the flood lights.

At Camaro, we eased into the cove and the driver pulled up as close to the shore as possible without risking damage to the boat. The water smelled like old fish and moss. We walked down a short plank that the driver lowered, but it didn't quite reach the shore. As we stepped off the plank and managed our way through shallow water, I was glad I had worn my waterproof leather combat boots instead of my hiking boots. The combat boots were heavy, but they were tall enough that they kept out the icy water. Wet, frozen feet aren't just painful, they could end up causing a thermal injury.

Twilight had done well in picking our insertion point. This cliff was steep, but I was confident we could climb it.

I loved having Twilight on the op. We leaned on him as our subject matter expert for all land navigation matters. He was an Army veteran and a former police officer. He was solid. He had an underlying drive to defy the norm and think outside the box. Even when he had been a new operator on our team, I could tell I'd want to have him on my sniper team someday. His maverick ideas combined with his subtle personality and devotion to teamwork were what I was looking for in a sniper. He had earned his nickname when he casually told the team that his wife loved fictional romance books about vampires and werewolves, but he really screwed up when he confided in us that he enjoyed them as well.

"You know, guys, they aren't that bad. You should read one."

"Okay, Twilight." And like that, he had a call sign.

I heard Twilight's voice over my Peltor headset say, "TOC, this is QRF. We're passing checkpoint Camaro."

OFFSET INFIL

Climbing the cliff was a challenge. We had flipped up our NODs and activated dim red light on our helmets. The NODs provided an acceptable field of view and depth perception, but not like the good old Mark-1 eyeball. We were still 800 meters from the objective and there was no danger of anyone seeing the small red lights moving up the cliff. I had practiced an 800-meter march in full kit and with a full pack a few days earlier to prep for the mission, but I didn't do the 1,000-foot climb beforehand due to a lack of cliffs anywhere near where I lived. I was breathing pretty hard, and everyone else was as well. I made a point to not use any nearby trees as support. Even a slight pull on a branch can make the top of a tree sway slightly. It's not noticeable from the ground, but to an observer in an elevated position, it's an indicator that there are humans in the woods.

Once we got to the top of the cliff, we briefly formed a small outward-facing circle and stood silently. We all listened to the environment and caught our breath. The only sounds in the forest were owls and tree frogs. Then everyone faced inward. I gestured a vertical line from my forehead to my chin with my left index finger to indicate we move in a single line formation, known as "ranger file." Ranger file was the best way to move through the thick cedar forest while keeping noise levels to a minimum. K2 would take up rear-security. Twilight would stay in the middle of the stack as our navigator. I held my hand up with my fingers together and palm out, and waved it forwards twice, signaling to "move out."

The bullfrogs serenaded us with their shameless mating songs as we did our best to avoid snapping branches with each step. About thirty minutes into the hike, I heard Twilight's hushed voice over my headset. "TOC, QRF is passing checkpoint Corvette." *Halfway there. Keep your head up and eyes out.* The closer we got, the more we would focus our efforts on being silent. The little red lights on our helmets went out and we flipped our NODs down.

About 45 minutes into the hike, my GPS indicated that we were nearing the Objective Rally Point: Trans Am. Sleet was covering my clear-lens glasses. I could have just taken them off, but in the darkness of the forest, I didn't want to risk getting stabbed in the eye with a stray tree branch. The QRF would stay at Trans Am where they were out of the subject's line of sight, but equidistant from each sniper team in case they were called to back us up. The four of us split into two sniper/observer teams. I headed left with Febreze, and Dangler headed right with K2.

Through my NODs, I could see a faint light strobing silently overhead. I clicked the mic button on my chest.

"Voodoo One, this is Sierra One. Are you on-station?" asking if he was in a circular orbit over the objective.

"Ahhhhfirmative, Sierra One."

"I'm glad to have our guardian angel overhead tonight."

"I'm just happy to be here and serve my country, sir," Al said with just the perfect combination of conviction and sarcasm. I knew that he must have been at the hangar at 3 a.m. that morning to complete his preflight. We both were functioning on minimal sleep. Big Al realized yesterday that his home airport would be socked in with bad weather this morning, so he preemptively flew the plane to Arkansas last night because the forecast for that airport was clear for the next morning. He got a room at a local hotel to get a little sleep, and launched again before the sun was up. It was a lot of effort for him to make sure that we had air cover for the op. It would have been easy for him to just scrub the mission for weather, but Big Al was a cut above the rest.

As Febreze and I got close to our pre-designated hide location, also known as a Final Observation Point (FOP), I flipped up my NODs and checked my GPS. I felt confident that I was about 200 yards from the objective and about 100 yards from my FOP. I pressed the button on the upper left part of my armor and keyed up my radio.

"Voodoo One, Sierra One."

"Go ahead, Sierra One," Al said.

"I show my position 200 yards from the objective, bearing two-one-five degrees. Do you have my beacon?" I flipped my NODs back down to make sure the beacon on my shoulder was flashing an infrared pulse that was visible to night vision, but not to the naked eye.

"Ahhh...copy. I got you," Al said.

"Voodoo, can you light it up?" I wanted more confirmation that we were headed to the right place.

"Arming laser."

Suddenly an enormous flood-light of infrared laser erupted silently from the sky. For every other human and animal in the forest, it was perfectly dark. Through my NODs, I could see the house lit up in spectacular detail for just a few seconds.

I took mental note of the layout of the structure before the laser switched off. That was an extra bit of situational awareness I needed before we moved to our final hide position. "Tally ho, Voodoo. Thank you, sir."

Big Al clicked his mic twice, indicating that he heard me. I checked my GPS and headed to our final hide location.

SURVEILLANCE

Once we got within twenty-five yards of the FOP, I dropped my ruck at a cache point. I needed to lower my profile to move to the hide. There's not much use in low-crawling when you have a pack on your back giving away your position. I slipped my bolt-action rifle from the opening at the top of the bag and quietly attached a PVS-22 night-vision device forward of my scope. I also attached a small IR flood light to help illuminate the structure. I took off my helmet and NODs and put a ghillie hat over my headset. My ghillie hat was a wide-brimmed jungle hat that

I had camouflaged with netting and jute that matched the local vegetation colors. The netting was long enough to completely cover my face. I had a small piece of panty hose to put over the objective lens of my scope to prevent the sun from reflecting off of it and giving away our position once the sun came up.

We silently crawled toward the objective until we got to a small ridge where we could hide behind defilade. With each movement, I was actively trying to keep snow from going up my sleeve. Once the sun came up, the only part of us that would be visible would be my camouflaged face under camouflage netting. For now, we still had the darkness for cover.

The strap from the downlink was digging into my wrist and annoying me. I slipped it off my left wrist and pushed it into the snow bank in front of me so I could see it when I needed it. *I can't forget to get that back. Radio would kill me.*

The downlink flickered blue for a second when it lost signal and then lit back up with a black and white bird's-eye-view of the objective provided by the plane above, orbiting in a slow silent circle. I could see what Voodoo's FLIR camera saw. I was the little white dot in the middle of the screen and the dot to my right was Febreze.

I quietly lowered my bipod, turned on my night vision scope, and chambered a cartridge. I put a camouflaged bag filled with rice under my buttstock and positioned it so that my rifle pointed at the objective without any input from me. I put a second bag filled with navy beans under my right elbow. I had made these bags when I was a new sniper and was fond of them even though store-bought options filled with more durable, synthetic material were available. I liked to joke that the new ones were nice, but my "retro" shooting bags would make a nice meal in a pinch.

Febreze had his M4 carbine with an EOTech holographic sight. He was looking forward at the objective instinctively.

"I need you watching our six." I whispered. He wrinkled his nose. No one likes to cover rear security. I said, "If a gunfight breaks out this way…" and I pointed at the cabin, "…you'll know." And I gave him a smile.

I looked through my Laser Range Finder, put the crosshairs on the house, and pushed the button to fire an invisible laser at the cabin. The display read "96 yards." I grabbed my compass and pointed it to the cabin as well. The dial rested on 32 degrees. *North-north-east. That makes sense.* I added 180 degrees to the heading to get my bearing back so it would make sense to the TOC when I called in my position.

I pressed the bottom button on my comms box to use the radio that was pre-set for the TOC-Net frequency. Radio and the other ETs had set up a thirty-foot antenna at the TOC with a repeater and had put another repeater in the back of Voodoo. When I transmitted on this radio, the signal was amplified and re-transmitted from either the fixed or the orbiting repeater, and I was glad for the redundancy. I had been on ops in which I had lost both radio signal and cell phone signal, and it could have been catastrophic. If I didn't have my rifle, but still had a functioning radio, at least I could still call out threats to my teammates and be useful from my hide. The radios on my back were my primary weapons right now.

"TOC, Sierra One." I could see my breath in the cold night air as the words came out.

"Go for TOC." I could hear chatter in the background noise. I was confident that the Tactical Operations Center crew was sipping hot coffee and eating donuts. *I cannot wait for breakfast. I don't care if I get stared at for my "guy-liner," I need some bacon and eggs.*

"Sierra One is in position. We are ninety-six yards from the objective bearing two-one-two degrees."

"TOC copies."

"TOC, Sierra Two," Dangler was whispering in his headset mic a couple hundred yards to my right.

"Go ahead, Sierra Two." *I swear it sounds like TOC's sipping hot coffee right now.* I started wiggling my toes in my boots to stave off the bitter cold, but it wasn't working.

Dangler's voice came across my headset. "Sierra Two is ninety-four yards from the objective bearing one-zero-zero degrees."

"TOC copies."

Back at the staging area, Lil Toe was watching his downlink too, and he could see us moving into position through the same thermal video feed that I had next to me. He told me later that it felt like we were alone in the mountains about to head to the Unabomber's remote cabin.

A light came on in one window at the back of the cabin. With my left hand, I clicked the top button on the comms box on my chest rig. "OC-1, be advised, I have a light on in black-bravo-two. He's up." I turned the magnification up on my rifle scope and looked intently to see if I could identify the subject from my location.

"Copy, Sierra One." Lil Toe answered into the left ear of my headset on Fight-Net. Keep us advised if –"

Before Lil Toe could finish, I heard Big Al's voice in my right ear over the air-to-ground frequency, urgently saying, "Sierra One, Voodoo. I have multiple heat signatures inbound on your position." My heartbeat picked up a little. I pulled my cheek off my rifle stock and looked at the downlink nestled in the snow. *Crap. Are those his dogs?*

Whatever they were, they were moving directly towards us, and fast. I glanced over my shoulder – Febreze was still watching our six – then pressed the button to release the retention device on my .45-caliber Glock with my right hand. With my left hand, I clicked the bottom button on my comms box and said, "Copy Voodoo. Thanks for the heads up. We have a tally on the downlink." Then I clicked the top button on my comms box. "QRF, this is Sierra One. We have hotspots on the FLIR moving directly to our position. We might need a hand."

"Say the word," Twilight answered over Fight-Net. He sounded calm but eager.

I slowed my breathing as I silently slipped my .45 out of the holster. If we went kinetic now, the assaulters would lose the element of surprise, and the subject would have time to prepare a defense.

Maybe they're just some spooked deer. Too fast to be human. Deer don't typically run TOWARD people, though. And I'm not about to get torn up by a bomb-maker's guard dogs.

I made eye contact with Febreze with my eyebrows raised as if to say, "Don't move a muscle." I unconsciously held my breath. I held my head still, but my eyes shuttled right and left, looking for a threat.

Silence. I willed myself to stop scanning and look at the downlink. In the black expanse of a cold forest, two white-hot signatures had changed course and were moving west – away from us. I quietly put my pistol back in my holster.

Big Al's voice interrupted the silence, "Sierra One, looks like those bogies are moving away from you. We will keep you posted." I got back on my rifle and pressed my cheek back onto the cold stock. Febreze had my six, the QRF was behind me, and Big Al was my eye in the sky. I knew they were looking after me, so I could get back on my rifle and look after the assaulters when they arrived. I scanned each window of the house one more time, waiting, daring the bomb-maker to show himself.

OVERWATCH

My ears perked up as a whine echoed through the woods – the big diesel from the MRAP, crawling its way up the road.

Lil Toe called out, "Sierra Units, we are one mike out." I dialed back my scope's magnification to get a bigger field of view and scanned every detail of the objective. *No surprises.* "Sierra One copies. There's no new activity at the objective."

This is where the sexy part of action movies contrasts boldly with our typical mission. There was a very real chance this subject had bombs and firearms. The Hollywood portrayal of kicking in the door and somersaulting through the threshold was the furthest from our *modus operandi*. No one needed to get shot or blown up, not even the bad guy. We brought overwhelming force and numbers to a potential fight, and the usual result was that no one wanted to fight us. Surrender was the norm.

The MRAP hissed and clacked as its air brakes activated and released. An operator was in the top hatch with a rifle pointed at the cabin. The PA squealed and then went quiet before a voice boomed out, "This is the FBI. We have a search warrant. Come out of the house with your hands up." We had "set isolation." Between the snipers, the QRF, the armored vehicle, and our aircraft, there was no chance of escape. Now we settled in for the potentially long wait.

But the front door opened, and a face peeked out from around the door. I had seen that expression before. He did not want this fight. Then he simply stepped out with his hands in the air.

"Turn around and walk backward to the sound of my voice," came the PA voice.

The subject walked back to the MRAP as directed and was met by an arrest team. Vulcher cuffed him and patted him down for weapons. Then he put his hand on top of the bomb-maker's head to keep him from hitting it on the doorframe of the sheriff's patrol car as he seated him in the back.

Well, that was easy. Yet another anticlimactic ending.

The team still had to clear the structure for people who might not have come out. From my hide, I watched them stack on the door and prepare to methodically clear the house using CQB (Close Quarters Battle) techniques. It was always assumed that someone was still hiding in the house. They were moving carefully and methodically, watching for trip wires or IEDs (Improvised Explosive Devices).

"First floor clear, heading to second deck," an assaulter called on Fight-Net. As a rule, no supervisors spoke on Fight-Net. It was just for critical traffic for the team on the ground and in harm's way. If a supervisor needed an important update, someone would call it out on the TOC-Net duplex channel.

I shifted my rifle up to the second-floor windows. As each head came into view through the windows, I looked for the friend-or-foe glint patch or flag on the side of his helmet. I scanned ahead of each room to look for movement. If I saw something out of the ordinary, I would halt the assault team with a radio command. If necessary, I would engage any hostiles from my hide to protect my teammates. As the team cleared through the first room on the second floor, I scanned ahead to the second room. "Sierra One, OC-1. Nothing seen in Black Bravo Two." I used our discreet code to let him know to which window I was referring. The assaulters entered the next room and cleared it.

Lil Toe's voice came over my headset. "All clear. Starting secondary searches. Sierras, collapse on the objective. We could use the help." I was suddenly aware of how cold I was, and I could feel myself start to shiver for the first time since we had boarded the boat hours ago. I put my rifle on safe, went back to get my ruck cached in the snow, and we hiked up to the cabin. In the distance, an ASAC (Assistant Special Agent in Charge) was standing outside the perimeter looking on, his hands in his armpits. There was always an ASAC on scene when SWAT was activated, but they rarely had any SWAT training or experience. He got paid a lot more than me, but I wouldn't trade places with him for any amount of cash.

"All units, Sierra One approaching the objective from the south. Blue, blue," I said. Febreze and I emerged from the snow-covered forest and walked up to the cabin. I didn't want to get shot by my own teammates by surprising them as we approached. Dangler was right behind us. His voice came over my headset, "Sierra Two approaching from the east. Blue, blue." I made eye contact with teammates who were holding security on the perimeter of the cabin.

I dropped my ruck by the MRAP and walked up to one of the breachers as he was putting away a ram, a Halligan tool, and other breaching gear.

"What? No breachers needed again? You gotta start earning your keep." I tried to keep a straight face.

Without looking up, he gestured to his ram, "Don't need lasers and range finders to make this work, sniper diva."

"Yeah. So simple a big, dumb ape could use it," I replied.

"Hope you enjoyed your nap out there, while we were up here where the bad guys are."

We both laughed. The sniper/breacher rivalry was a fake one, but it was a fire I constantly stoked.

I walked up to Lil Toe. Suddenly the adrenaline was gone, and I was jonesing for a hot cup of coffee. Still, I asked, "Where do you need me?"

"Jump in with ERT and help out where you can," he replied. I had been on ERT (Evidence Response Team) for six years during the same six years I was a new SWAT operator. I had applied for SWAT and ERT at the same time, not expecting to get accepted to both. Everyone on ERT was a friend of mine. They were Special Agents, Evidence Technicians, Rotors, Analysts, and Secretaries, but they had all gone to ERT school for at least two weeks and learned the basic skill set of the crime scene investigator. Everyone knew phenolphthalein presumptive blood tests, how to dust for prints, cyanoacrylate print lifting, crime scopes, forensic photography, and more. Without that training under my belt, I would have just been in the way during the search.

I found a group searching the upper floor and jumped in. The search warrant ordered us to look for evidence of the bombs, so we scoured through every nook and cranny in the cabin. One of the ERT techs opened a drawer, gestured to a heavy-duty stapler and a box of extra staples and glanced at me. "Stay right there," I said. "Let's photograph this in place." I called downstairs to the lead photographer at the ECP (Evidence Control Point). "Can y'all come up here? This needs to be logged," I said. *That's a great find, right there. I hope that's the one.* I patted him on the shoulder. "Nice work."

Since they did not pay for us to stay at hotels after one-day missions, that meant I needed to load up on a lot of coffee to make the drive back home. I walked down the stairs and out of the cabin. I looked over at Lil Toe as we walked out of the house and on to the front porch. I made a cheesy smile at Lil Toe and quoted from *Wayne's World*, "Almost toooooo easy." Lil Toe looked back with the smallest hint of grin and said, "Yeah. Not bad."

It dawned on me that after all that mission planning, we hadn't specified how the sniper team was supposed to get all the way back to the south side of the lake to retrieve our Bu-Cars (the FBI nickname for official duty Bureau take-home cars).

As Lil Toe started to walk toward his vehicle in the driveway, I said, "Hey, can we hitch a ride back to the rally point?"

Without looking back, he said, "Hell, no. You guys hiked in. You can hike out." After a second to absorb the sarcasm, I burst out laughing.

Once I was back at my Bu-Car, I hastily crammed all my gear into the back of my truck. I knew I would spend several hours the next day cleaning, reorganizing, and checking every item. For now, I was ready to get home, see my family, and sleep.

CASE CLOSED

After the blur of the next few weeks, I was surprised when I got a call from a colleague about the bombing investigation. The FBI laboratory had analyzed the stapler we had seized from the bomber's house, and they had finalized a forensic evidence report. Every stapler imparts a unique mark on each staple – akin to a human fingerprint or the primer indentation on a cartridge case. The microscopic mark that the stapler from the cabin made on staples was identical to the marks found on six of the seven staples recovered from the homemade bomb. What's more, the lab had confirmed that the elemental composition of the staples we found in the residence matched those from a disrupted bomb. Of all the ways a bomber can get caught, who would have thought that the stapler and staples would be the evidence that sent him to be the newest resident in a federal penitentiary.

I hung up the phone and leaned back into my seat. That was some of the best news I'd heard in a while. Dozens of hours of preparation had gone into that SWAT operation, and hundreds more hours were spent on investigation and analysis. Plus, the team had been on the road a lot the last two months to pull Dignitary Protection for the FBI Director and for the U.S. Attorney General. Those ops involved long days of wearing suits, waking up long before the dignitary, and going to bed long after him.

The next several days were going to be extra busy too. I had meetings with local police, SWAT training, firearms training, and surveillance assistance for the Joint Terrorism Task Force. Equally importantly, I had a lunch date with my wife, a parent/teacher conference, and Mackey's wedding. Plus, I had to fly back to Quantico for a week-long routine recertification as a firearms instructor, during which I would miss my wife's and my mother's birthdays. On top of that, I had to get prepared for an upcoming medical mission trip to Uganda with my church. This job never stopped, and it would eat you alive if you let it. You had to savor the victories.

And the victories were never mine alone. Every successful arrest served, fugitive caught, terrorist plot thwarted – we did as a team. I couldn't have asked for a better group of teammates. Each guy was extremely bright, driven, and would give you the shirt off of his back. They were all high-achievers in their last jobs and many of them had taken a pay cut and Transfer Orders that took them away from their hometowns and extended families when they joined the Bureau.

In addition to being great teammates, they were great friends. Mackey and I and our wives endured the toddler "terrible twos" together. We all attended Dangler's father's funeral. We sent gifts and visited K2's son after he broke his leg. We all hung Christmas lights on Wylie's house when the FBI sent him to Afghanistan, embedded in a military combat unit for three months. Together we spent a day moving Twilight's kids' swing set when he got a new house. Last year, I shoved my fingers deep into Nestle's upper thigh muscle to stop him from bleeding out from a bullet that had cut through his armor. The team sent books and toys to Vulcher's son when he was hospitalized for a suspected brain tumor. Vulcher's son and Twilight's daughter would grow up together and date in college. Frosty and JimBo flew to Walter Reed Hospital and spent two weeks with our teammate, White, after his Hummer flipped in Afghanistan. White broke both his legs, and his rifle barrel punched a hole into his armpit and out his back. We would all attend Febreze's infant daughter's funeral with our wives; wearing black suits with pistols under our jackets and holding pink balloons.

You know you are part of a special brotherhood when you walk through all of life together – the good and the bad. These guys were more than just teammates; they were wonderful dads and husbands. They were my family.

Chapter 7

DIGPRO

Mission #79
April 2015

"Hey, fellas. I'm gonna get started." The lights dimmed as the Detail leader started his briefing. I had my beat-up, light brown waterproof notebook out. It had "DIGPRO" written on the front. It was a sort of living-document that I used for notes from other Dignitary Protection missions. I had similar notebooks for aircraft hostage rescue, trauma care, and snipercraft.

I skimmed through my reminders to myself: *Brief the motors on our convoy ops.* That was an important one. The metro police motorcycle officers (known as "motors") didn't know that sometimes we swerved our SUVs into the left adjacent lane and straddled that lane when there were potential threats on the right shoulder. During the last DIGPRO, Frosty was bumping his SUV into the left lane to make distance from an abandoned car on the highway shoulder when a motor passed our motorcade on the left at 100 mph. After we stopped and unloaded our dignitary, the motor cop walked up, threw off his helmet, and tried to start a fistfight with Frosty. He thought we were careless drivers, but he didn't know our motorcade ops, and we had failed to explain it to him properly. Things like this are good to jot in my notebook.

"The Director will fly in tomorrow night," the Detail leader continued. He will visit the Oklahoma headquarters on Friday and be one of many dignitaries attending the ceremony for the twentieth anniversary of the bombing of the Murrah Building on Sunday."

There are lots of movies and TV shows that depict FBI Agents flying around on private jets, but the truth is we didn't get a Gulfstream G5 until after 9/11/2001. In fact, when I was a new Agent, FBI Director Freeh primarily flew on commercial flights for his official travel. Street Agents definitely didn't get to use the jet while working routine cases. There's not much reason for air travel for investigations in the FBI since we have fifty-six Field Offices, 400 Resident Agencies (RAs) and fifty overseas Legal Attache offices. If you needed something from a particular place, you just called the closest RA. They would always have contacts, from the mayor, to the sheriff, to school principals, and on a first-name basis.

The main reason for the purchase of our Gulfstream G5 jet was for Foreign Transfer of Custody (FTOC). That is, the FBI needed the capability to fly overseas, unrefueled, for rendition operations that brought terrorists back to the U.S. to stand trial. You don't want to stop in another country to refuel when you have a terrorist on board. It wasn't the Director's private jet, but he certainly made use of it for official travel when it wasn't being used for FTOC. Regardless of if it was the Director or a captured terrorist in the passenger seats, the pilots were usually gun-toting FBI Special Agents, assigned to full-time pilot duties, and ready to literally fly around the world.

There are full-time Detail Agents, and those guys had volunteered to transfer to D.C. to take the job. Many of them had been FBI SWAT operators when they were in the field, and that instant kinship made them easy to work with. This was the seventh time our team had been called up for a Dignitary Protection Detail. Someone needs to protect the Director and the Attorney General of the U.S. (AG); I just never thought about who it was until I joined the FBI. Working with the Detail was a bit like being a U.S. Secret Service agent for a few days. Truth be told, my team had gone to the local Secret Service on several occasions to ask for advice on how to do the job well. If they're responsible for protecting the past and present Presidents of the United States and their families, any tips they had for us were certainly worthwhile.

The dignitary protection details were fun for a few days, but I couldn't imagine doing it every day for a living. There's a lot of repetition, boredom, and stress for a job that was essentially thankless. I had done protective service details for foreign military dignitaries and four-star generals when I was a Special Agent in OSI, so I already had some experience with it before I joined the FBI. Everyone had a flavor for how they did DIGPRO, but it all had the same bones.

When the Director visited our area of responsibility, a few Detail Agents from D.C. flew in a couple days prior to his arrival to prepare and coordinate with my team. They didn't have the personnel numbers to do the entire job by themselves, and they weren't supposed to. Standard procedure was to activate Special Agents from the closest FBI SWAT team to complete the Detail package. In fact, we made up the majority of the DIGPRO team. We provided knowledge of the local area, local vehicles, medics, assaulters, drivers, and plain-clothes operators. We collected human intelligence, conducted surveillance, and provided liaison at each venue long before the dignitary and his traveling detail arrived.

I've been asked many times if I'd give my life to protect the FBI Director or the Attorney General. The answer is no. My teammates and I would make any would-be attacker pay a hefty price though, while the full-time Detail evacuated the dignitary. I usually chose to wear body armor under my suit on DIGPRO, but it was so I would have an advantage in a gunfight against an attacker, not because I planned on getting shot.

Pragmatically speaking, it's easier to replace a presidentially-appointed bureaucrat than an operator, anyway. The FBI Director usually has substantially less experience in the FBI than the average operator has in his own tradecraft, and the FBI could still run just fine until the President appointed the next person to fill the post. There hasn't been a director that had actually been a former Special Agent since Louis Freeh, and he retired shortly before the terror attacks of 9/11.

All that to say, we did our level best to protect our dignitary; there just wasn't any romantic notion that we were hoping to catch a bullet for a bureaucrat.

The Detail leader continued with the brief: "The Director will be in town to speak at the Oklahoma FBI headquarters building on Friday, then attend the Oklahoma City Murrah Building Memorial on the following Sunday. That puts him in a rare situation: he has a whole Saturday off, essentially stuck in Oklahoma. I rarely see the boss take a day off for anything, but he mentioned that, given the opportunity, he might try to get in a game of golf on Saturday. He mentioned some golf course near Stillwater." He noticed the look on my face when he mentioned the golf course. And my hometown.

"Carson's Creek or something?" he asked.

I nodded and smiled. "Karsten Creek," I corrected. "It's just outside Stillwater. That's my RA territory."

"Great! You know about the place?" he asked.

"Yeah. I don't golf, but I know the place. I eat lunch there sometimes with my wife. It's nice," I answered.

"GPS doesn't recognize the address and it's a two-hour round-trip. Could you help us out?" he asked, but his tone of voice didn't sound like he was asking.

"It's actually pretty hard to find. It's not well marked." I immediately regretted saying that. I had just given him a really good reason to rope me into working yet another Saturday.

"So, you can drive us there on Saturday, right?" he asked, still not in an asking tone.

I hated to be cooped up in the car for hours at a time, but it was hard to say no.

"Sure. No problem." I sighed. "I can make sure he gets there on Saturday." My wife and I had dinner plans with friends for Saturday. *She's not gonna be happy about this. But I can't say she's not used to it.*

"Actually, why don't you just drive the Director's limo the entire weekend?" the lead Detail Agent said. His voice was upbeat, like he had just come up with a convenient solution.

I said, "No. That's okay. You should find someone else. I live in Stillwater, and the limo will be in the metro area almost the entire trip. I can meet you at the golf course on Saturday to get you all set up. You'll want Agents from the metro area for drivers for Friday and Sunday."

He said, "No worries. We'll just put you in our hotel all weekend. That'll make things simple."

I nodded and strummed my fingers on the desk. This wasn't ideal, but I knew I'd ask for the same favor if I were in his shoes.

"So, Friday you'll drive him to the FBI building. Saturday, to the golf course. And Sunday, to the bombing memorial. Cool?" he asked. He seemed happy that I was agreeable.

Technically there's no "rank" in the FBI. Supervisors and managers got slightly higher salaries and nicer cars, but it wasn't like the military at all. There were no "direct orders." You could fight a decision if it was wrong or you felt strongly about it – you just had to have a legitimate reason and the stomach to put up with a ticked-off supervisor. I could refuse to help, but why would I? This guy was my equal pay grade, and he was being pretty cool. The plan made sense, even if it ruined my weekend. I would have hoped for a helpful Resident Agent if I were in his shoes.

"You bet. I got you," I said.

I had only been a driver on a couple occasions, and I did not enjoy it. I was usually in the convoy as a counter-assault team member. If not there, I usually acted as lead medic, sitting next to a suitcase-sized trauma kit in case the dignitary was injured.

If anyone ever asked about the job of a motorcade driver's responsibilities, he always got the same answer: "Drivers drive." You only have one job. One long, boring job. Never leave the truck. Always be ready to evacuate. Never turn the car off. And make sure you have a good battle-buddy to swap out with you when you need to go pee.

MOTORCADE

A dozen metro motor cops were gathered around their fleet of a dozen police Harley Davidson motorcycles in front of the airport. I always spent some time talking to the motors before a motorcade, just to build rapport and make new friends. It was early Friday morning, and we were prepping to pick up the Director at the airport. Coordination with the local airport had gone smoothly, and we had Agents in suits as well as in low-profile clothing all around the airfield. We had obtained the access code to the tarmac so we could drive the motorcade directly up to the jet. That way the Director could walk down the steps of the jet and directly into my limo in just a few seconds with minimal exposure to threats.

The motors were all casually telling stories and acting like it was just another day, but under the surface they were buzzing with excitement.

"Y'all collect patches?" I asked as I passed out FBI SWAT unit coins and patches to a few of them. One of the motors gave me one of his department patches, and I was excited to put it in a place of honor in my collection by my desk tomorrow.

Then one of the motors said, "It seems pretty stupid that we're just standing around at the airport three hours before the Director's arrival. It just seems like a waste of time, you know? Why not get here right before he lands?" he asked.

Okay. Be nice. This guy is just the grumpy one.

"I get it, man," I said. "It's a lot of hurry-up-and wait, but he's on a government jet, not a commercial plane. If they make good time, they aren't going to throttle back or circle to land. If they get a tailwind or the take-off time changes, they could land here an hour early. If no one was here to pick him up, there would be righteous hell to pay." I smiled at him and shrugged. "At least it's a beautiful day, and we're getting paid double time to drink coffee right now."

"You guys get double time pay for this?!" His eyes got big.

"I'm just messing around," I said. "We never get overtime. We probably won't even get comp time for the weekend."

"Still. Getting here an hour early would be plenty. Typical government," he said with a frown. I wanted to say something about how the metro police department was also "government," but that wouldn't help things right now. I could see his fellow motor cops rolling their eyes at him in my peripheral vision.

"So, the Detail leader said we really only need six bikes for this motorcade," I bluffed. "If you want to take off, it's no problem at all."

He stared at the ground and shook his head. "Nah, I'm good."

"No, seriously. Get out of here, brother. No one will mind." I nodded toward the airport exit.

One of his fellow motor cops had enough of it. "He's full of shit. He's been talking about this all week. He's not about to miss this. Plus, we get the rest of the day off after we get you guys to the hotel." We both laughed and Grumpy even managed a smile. Mission accomplished. I knew part of today's team a little better, and I had some new friends.

I heard the Detail leader's voice over my earpiece. "He's thirty minutes out. Line up the motorcade." We drove the Lead, Limo, and Follow vehicles onto the airport tarmac. Black-and-white metro police cruisers took up positions at the very front and rear, with Detail Agents in the passenger seats. Big black federal SUVs are sexy and all, but if you want to part the seas on the interstate at rush hour, black-and-whites and motors were the solution to any traffic problems. When people see black SUVs with police lights, they slow down to take photos. When they see a trooper or a metro cop, they get out of the way. The local police and troopers had no obligation to help us, and it was a huge favor to have them on the motorcade.

I did a last walk-around of my armored steed and slipped into the driver's seat. My earpiece was in my left ear with a small coiled pigtail connecting it to a wire that ran down my sleeve and into the radio on my belt. Belt real estate was limited, but I crammed my Glock 19, a spare magazine, a radio, handcuffs, and my badge on my belt. I had a tourniquet in my pocket, and I had moved my wallet and credentials to my suit coat pockets so I didn't have to sit on them for hours. In Bureau tradition, I kept my suit coat on in order to be "professional."

I carefully pulled up to the jet and parked as close as I could to the jet's stairs. I scooted my seat up as far as I could to make footroom for the Director, since he would sit directly behind me. As the back-right door opened, the roar of the jet's engines crescendoed. The Director slipped in the door and scooted across the bench seat until he was behind me. Then his staffer got in and sat to the Director's right.

"Thanks for driving me, Jeremy." He remembered my name – or at least asked someone to remind him of my name. *That was classy.*

"You're welcome, sir. Do you have enough legroom?" I asked while looking at the Director and his staffer in the rearview mirror.

The staffer kindly answered, "I'm good."

The Director looked at his staffer and said, "He means me, not you. It's my legroom that's important." They both laughed, and I cracked a smile. *He has a sense of humor. Maybe this will be an easy weekend.*

Speeding down the highway, I kept a half car-length between me and the Lead SUV. The Follow SUV was right on my tail as well. Predicting when to brake was critical. Big armored limos don't slow down fast, and playing bumper cars at these speeds was unacceptable. The black-and-whites in front and back had on their red-and-blue emergency lights. The motors would speed ahead and shut down intersections. Then they would catch up and race *machety-mach* past my limo to the next intersection. They were skilled motors for sure. And they were having a hell of a good time racing their Harleys around on the highway.

At the end of this organized chaos, one lonely SUV followed behind the last police car. It was filled with staffers and interns that were universally disappointed that they didn't get dignitary treatment or even get to stay in the convoy. The only car further back from the staffer car was the luggage van that was usually driven by one of the newest Agents in the division. If the staff or luggage cars get stuck at a light, they don't run it with us. They would wait for a green light and play catch up while trying not to get pulled over for speeding. It's tough to argue with a trooper that you have to race down the highway to deliver luggage.

"Broken down car, right side," the Lead SUV called out.

It might be a broken-down car, and it might be a roadside bomb with a shaped charge set to detonate as the limo passed. This was the way that Alfred Herrhausen, the Chairman of Deutsche Bank, and Rafic Hariri, the Prime Minister of Lebanon were assassinated – even while in an armored vehicle. We didn't have time to debate; that's why we had standard defensive procedures.

I instinctively eased my armored yacht-of-a-truck into the middle lane. The Lead and Follow SUVs stayed in the far-right lane and formed a protective shield with their vehicles.

"All clear," the Follow Agent called on the radio. The exit for the Oklahoma FBI HQ was just ahead.

EVERYONE GO TO BED

This was the second time I'd been on this Director's detail in just two months. In my career, I would conduct protective details for eight different AGs and four different Directors, and this guy was by far the most laid-back. After the Director's visit to our headquarters, we took him into town for a dinner meeting, then brought him back to the hotel to "put him to bed" for the night.

I was oddly impressed that the Director was staying at the simple Marriott across the street from the FBI field office for this visit. Most Directors and AGs stayed at the nicest hotel in the area, even if it was overpriced and an hour away from the venue. I admired that he wasn't high maintenance and hadn't spent taxpayer money for a luxury hotel. After all, he was working, not on vacation. After dinner, we escorted him to his room, and he wasn't expected to come out until early the next morning.

I decided to stop by and visit with the Command Post (CP) crew who were sitting in a secret room in the hotel. Local Special Agents, Electronics Technicians, and newer operators from our team used expensive gadgets to make sure no one tried to access our restricted area of the hotel. Every room in this hall was rented for us, so there was no reason for any visitors.

I liked to chat with the CP guys on shift because I knew what a tedious job it was, and I wanted to encourage them. They knew that I had escorted the Director to a five-star steakhouse while they were stuck in the CP all night, and they might have some fictitious notion that I was enjoying the creature comforts of being out in the field. I wanted them to know that once they moved up to the "big leagues" like us, they too could sit in suits in an SUV in a back alleyway for three hours, smelling delicious steak mixed with the smell of the adjacent dumpster, while eating a protein bar. I couldn't even get someone to bring me a cup of coffee while I sat in the limo. At least they had free coffee and decent snacks in the CP.

As we sat chatting in the CP room, one of the monitors squawked, indicating that a sensor had gone off down the hall on our floor. The screen showed two white males walking toward our hallway. Nilla and I stepped out into the hall and stood shoulder-to-shoulder as we faced a couple of staggering fellows who reeked of booze.

"How come y'all need all this security? Is Puff Daddy in town?" The bigger one slurred. It was a delicate balance of respecting a drunk redneck's right to free speech, protecting the Director, and my desire not to take crap off someone who was looking for a fight. The big one tried to slip past me, and I blocked him with my shoulder. The little drunk guy said, "Moovit, man. My room's down there."

"No. It's not," I said. "Go home," and I nodded toward the elevator.

"What?! Are you gonna arrest me? Huh?! Wanna *make* me go home?" He made fists with both hands. His friend wasn't quite so aggressive, but he still squared off.

Nilla Dean wasn't a gigantic guy, but he was supremely fit and appropriately confident in his ability to hurt bad men. He took a step forward toward the two much larger men and said, "Try it." He tapped his finger to his own chin. "Go ahead. See how it ends."

There was a long tense moment.

He said, "No? You sure?"

Nilla had so much composure and confidence that I was a little scared of him myself. Heck, I was slightly sexually attracted to him at that moment. That was badass stuff. The two rednecks knew they were in for pain if they took that swing, so they hung their heads and wandered down the hallway back to the elevator. *No more mischief tonight, please.* I checked my pocket to make sure I had my car keys and went back into the CP room to grab my suit coat.

As I slipped on my jacket just inside the CP, suddenly, one of the surveillance cameras showed the Director stepping out of his room, unannounced. *This night will never end.*

"He's up!" one of the CP Agents called out, as surprised as any of us. I looked down the hall. Looming at 6'8", the Director was nearly a foot taller than me. Earlier today I had noted that if I walked directly up to him, his tie tack would have hit my nose. My own tie was loose and the top button on my white dress shirt was undone. I was off-duty now and I hoped he didn't mind my casual dress – and hadn't overheard that confrontation.

He looked at me and smiled. "I have a secret mission for you, if you're willing to accept it." Then his face hardened up in facetious seriousness. He was holding two tiny complimentary shower bottles in his hand and handed them to me.

"Yes, sir." I wasn't sure whether to laugh or act serious as well.

"These are two bottles of conditioner. I'd like for one of them to be a shampoo." And he let a smile crack.

I wasn't his personal assistant, and my shift was long over, but I supposed it wasn't too much to ask to get the man who supervised 35,000 FBI employees a tiny bottle of shampoo. I started to laugh. "Of course, sir. I'll see what we can do."

"UBUR"

I headed out of my hotel room around 4 a.m. It was Saturday – day two of three. I was wearing a dark gray suit with a crisply starched white shirt and a subtle blue tie. I also had Captain America cufflinks on, just to spice things up, though I doubted anyone would notice. I had my earpiece hanging over my collar. It gave me a headache if I kept it in for more than a few hours, but the earpiece lent a surprising air of authority. When I got coffee at the hotel, long before breakfast was served, I saw some hotel employees staring at me, specifically at my earpiece.

"Breakfast isn't open yet, but you can go in the kitchen and get some cereal," one of them said.

"That's so nice! I really appreciate that." I said.

She gestured at my earpiece. "You look like someone important."

"Oh, I assure you. I'm not." I laughed. "But I'll take you up on that bowl of cereal."

There was nothing glorious about this. I might have been just a chauffeur on this day, but I was a chauffeur that made good money and got to work righteous federal cases. Besides, sometimes you should relish the easy days – even if they are Saturdays when you could be at home.

After I ate, I drove my Bu-Car (my take-home car), a well-worn Ford SUV, to the undisclosed location where we hid the 11,000-pound armored Suburban. That big beast of a truck is what we referred to as the "limo." I made certain it was cleaned, fueled up, bomb-checked and ready for the Director. Driving the armored Suburban "limo" was like driving a tank with a dying engine and ruined brakes. The limo didn't like accelerating, but it really hated stopping. You had to plan ahead to get it stopped, and it needed new brake shoes regularly. The ballistic windows were over an inch thick and only rolled down a few inches. They weighed quite a bit and it was comical how slowly they rolled back up. I noticed when I did my walk around of the limo that on the driver's door, someone had pulled a few of the letters off the "SUBURBAN" logo. Now it just read, "UBUR." *That's about right: I'm an Uber driver with a gun, driving a shiny black tank with bad brakes.*

GOLF SHOES AND MACHINE GUNS

"Coming out," the lead Detail Agent's voice whispered in my earpiece. I had been sitting outside the rear exit of the hotel with the engine running for almost an hour.

An Agent opened the back door, and the Director climbed into the limo. For the first time, it was just the Director with no staffers on his coattails. We sat quietly while I drove him to the golf course about an hour away. No motors zooming by. No black-and-whites. No lights, no sirens. He seemed like he was enjoying the silence, so I just sat in the quiet with him for the drive.

When we got to the golf course, some of our Detail Agents walked with him, wearing polos and khaki pants with golf bags of their own. The best way to do a protective detail in my opinion is to have as little pomp and circumstance as possible. There was no one at the golf course. No paparazzi, nothing. It was ideal.

I first learned how to conduct dignitary protection while I was a Special Agent in the Air Force Office of Special Investigations (OSI). We would practice in downtown Washington D.C., with one of the instructors serving as a make-believe dignitary. We drove large SUVs and wore black suits as one might expect. When we arrived at our first stop in Martha's Vineyard, I remember people swarming the limousine to try to take pictures of our instructor, simply because he was a man in a black suit, surrounded by other men in black suits and sunglasses. People took photographs of him and asked for his autograph. He wasn't even a real dignitary, but people can't help but be drawn to the appearance of someone rich or famous or powerful. So, it was a good idea to now just have the Director golfing with a few casually-dressed Agents and enjoying the day.

After about nine holes, dark clouds started to roll in and it started to sprinkle. I sat in the limousine, listening to audiobooks and answering emails on my phone. When the Detail Agents would come back to the motorcade for breaks, I'd ask one of them to grab me a cup of coffee, but no one did. I didn't have a cup holder, anyway. *Eleven thousand pounds of badass armored SUV and they didn't think to include a cup holder for the driver.*

I heard the lead Detail Agent say in my earpiece, "It looks like a storm is coming in. We're going to head back." I put the limo in drive and drove up to the main entrance of the club house. There, a Detail Agent peeked out the back door and made eye contact with me. The Lead and Follow SUVs took their places in the motorcade line up. A second later, the club house doors opened, and two Agents came out with the Director.

Rain was already splattering across my bullet-proof windshield. The Detail Agent opened the back door, put the Director in my limousine, and shut the door behind him. I prepared for a nice quiet drive back to the hotel. I figured the Director regularly got bugged with mundane questions, and he deserved to have at least a half day in peace and quiet. Plus, there was an unspoken rule not to address the Director first when you drive the limo.

He surprised me and broke the silence by saying, "Jeremy, this is a really nice golf course. Have you ever golfed here?"

I still can't believe he remembers my name. "No, sir," I answered. "I've never golfed here, but my wife and I get lunch here sometimes. It's beautiful scenery and the food is pretty good. I'm not much of a golfer, though. I've been chased off of more than one course." *That was a long answer. I could have just said, "no, sir." Aren't we chatty.*

"Really?" he asked.

"Apparently, it's frowned upon to drive your cart onto the green, or to take twelve putts." I could see him smile as I looked in the rearview mirror and was glad that he understood the humor of it.

He said, "I can't believe you live here and you haven't golfed here. You should try it sometime."

"I don't know much about golf, but I know one joke about it," I said. As soon as the words came out of my mouth, I fully regretted them. I curled my toes and held my breath, hoping that somehow I could take back what I just said. I had painted myself into a proverbial corner, and now I had to tell a mediocre joke.

"Let's hear it," he said.

I took a long breath as I accelerated down the on-ramp and merged onto the highway. "It goes like this…" I began.

> "A father and son were golfing when a thunderstorm rolled in and it started to rain. The son said to his father, 'We need to get inside before we get struck by lightning.'
>
> "The father answered, 'No, son. I know a surefire way to never get hit by lightning.'" Then the father pulled out a one-iron and held it straight in the air. The son said, 'Dad! What are you doing!?'

"The father, holding the golf club over his head, defiantly said, 'I'm fine, son. Not even God can hit a one iron.'"

I blinked hard and looked in the rearview mirror. I saw to my horror a complete blank stare on the Director's face. *I told a really dumb joke to the Director of the Federal Bureau of Investigation. I'm an idiot.* Worse yet, I told a semi-blasphemous joke to a man who also happened to have a degree in religion. He just made a humming noise to himself and said. "Huh. Okay, okay."

It's a good thing I wasn't interested in climbing the promotion ladder in the FBI.

DAY THREE

Preparations for the twentieth anniversary of the Oklahoma City Murrah bombing were hectic. There would be a lot of dignitaries in one place at one time. It wasn't lost on us that we were all packed into an open public space to pay respects on the same soil where a domestic terrorist had detonated a bomb and murdered 168 innocent people. My team was protecting the FBI Director and our sister team from Kansas City was in town to protect the United States Attorney General. The current U.S. President and a former U.S. President were both attending, and the Secret Service had multiple teams at work protecting them. When we arrived at the event, the motorcade of armored black Suburbans seemed to go on for miles.

We usually worked very well with the Secret Service. We had planned to put snipers on the rooftops in camouflage uniforms for this event, but the Secret Service asked us to wear blue raid jackets so they could identify us quickly. When the Secret Service is protecting the President and the FBI is protecting the Attorney General and the FBI Director, the Secret Service can "pull rank" after a fashion. That's about the only time the Secret Service can tell the FBI what to do, and we respected it. Their counter-sniper teams wanted absolute identification of who the good guys and bad guys were, so our sniper team agreed to wear blue jackets with yellow FBI letters. It looked sort of silly to be on the roof with a rifle and big blue raid jacket, but we were all on the same team, and we intended to play nice with our brothers and sisters in the Secret Service.

We were waiting our turn to drop off our dignity near the outdoor stage at the ceremony. As I sat in the long, meandering queue of black limos, I looked up and saw one of my sniper teammates on the roof looking through a spotting scope. *I wish I was on the roof with my boys right now and not driving this behemoth SUV. I bet they even have coffee up there.*

WHEELS UP IN TEN

This DIGPRO op was winding to a close, and I was ready to see everyone climb on that expensive jet and leave my territory so I could take off my tie and get out of this driver's seat. The last stop was the general aviation terminal at the airport. The motorcade pulled up to the airport tarmac gate, and it opened for us. I slowly pulled the limo up to the staircase of the waiting jet. The engines were roaring, even at idle power. All the Agents got out of the motorcade and set up defensive positions facing in all directions, taking one last scan of the area before the Director stepped out. It was just the Director and I alone in the limo again.

"Thanks again for driving, Jeremy," the Director said as he reached forward from the backseat. He gave me a firm handshake and I felt a coin in my palm as we shook hands. I took a quick look. It was his coveted personal coin. *I guess my jokes aren't that bad, after all.*

"It was my pleasure, sir," I said genuinely.

The door opened and the Director walked up to the side of the jet. Detail Agents were hastily putting their rifles into rifle cases and moving luggage onto the jet. In the meantime, per tradition, the Director took photos with the local Agents and police as the jet engines screeched away. Then he climbed up the stairs, and the jet taxied out to the runway. We waited on the tarmac until the plane was wheels up and then we waited another thirty minutes in case the plane should come back unexpectedly. After that, I gave the limo keys to a local Agent, and hopped a ride with a teammate back to the hotel to pack up.

I have a lot of respect for the Agents who volunteer for full-time Detail. They live hectic lives. Initially I envied that they got to polish their craft to perfection by daily repetition. Get up, protect the boss, go to bed, repeat. No worries about yesterday, or a week from now, just tomorrow.

As a street Agent working thirty cases, my life was distinctly different. I had so much variety each day that I worried I'd get an ulcer. Each day brought new interviews and evidence. Each new lead brought me closer to solving a case, but it also brought new questions about what steps I would need to do next. There was never enough time, and I always felt rushed. It didn't matter if I was on vacation or on sick leave; my cases banged around in my head all day, every day. It wasn't unusual for me to wake up in the middle of the night to write down an idea for solving an investigation.

Even so, if I could choose anything, I'd choose my current path: to be a street Agent, actively running my own investigations, and serving on the SWAT team. I could see the attraction of being on the Detail, though. It was a lot of excitement for a few days, but I was glad my part in it was over for now.

As I packed up my hotel room, I saw my light brown waterproof notebook labeled, DIGPRO, on the nightstand. This was the ideal time to jot down some notes for how to better execute the next detail:

- *Don't mention you're familiar with the venue, unless you want to get selected as the limo driver.*
- *Keep an energy drink with a screw-top lid in the front seat with you if you are a driver. Hot coffee will never be available.*
- *Never offer to tell the Director a heretical joke about golfing.*

Chapter 8
Sanctioned Killing
Mission #95
June 2017

The advertisement read: *"For production overseas, looking for talent, 30-45 years old. Doesn't have to be a professional actor. Required: Creative, outgoing and friendly, positive personality, boldness and bravery (some stunts may seem risky, although they are completely safe). You must not disclose any information related to this plot to ANYONE under ANY CIRCUMSTANCES."*

The author of the Craigslist ad, Samantha, got a response from Tommy, a.k.a. Tammy, a transgender man who was an intact male with breast implants. Tommy thought the advertisement was a euphemism for some sexy debauchery. He replied that he was "down for anything" and had "a nice set of double Ds." Despite the misunderstanding, Samantha agreed to meet with Tommy in person.

Samantha nervously ran her fingers through her brunette hair, but composed herself as he walked into the room. He couldn't hide his nerves as well as she did as he made an awkward smile.

"Have a seat," she instructed him as she gestured to a chair by the table. There was nothing noteworthy about Samantha other than that she was pregnant. She was an average-looking white woman in her late thirties. "Nice to meet you. I'm Dana," she said. Samantha had no intention of using her real name. She opened up her laptop to a presentation entitled, "Operation Insecticide." She had rehearsed a detailed presentation with instructions on what he was being asked to do, what he should pack, how he would travel, where he would stay, and how he should communicate.

"This is a secret assignment. I'm an Israeli intelligence officer, and I'm commissioning you to travel to Israel to kill an ISIS operative." Tommy struggled to hide his shock at the words she had just spoken, but he was oddly intrigued and decided to hear her out.

"This is a sanctioned killing," she said. "You will be doing the world a favor. You'll be paid $4,000 if you're successful, plus $1,000 for your travel expenses. The ISIS terrorist is a cab driver, and your job is to flag down his cab for rides over the course of several days. Learn his favorite coffee on the first ride, and bring him a cup on the next ride. Make sure you put a little of this in his coffee on the second ride." She handed him a written list of his mission objectives along with a plastic baggie with a powdery substance in it. "Be very careful with that. Use gloves when you open it. It's a lethal toxin called ricin." Tommy agreed to carry out the secret mission and headed home. There, he showed his roommates the baggie and the instructions. One of his roommates immediately threw the baggie into the trash and another one secretly called the FBI.

The Oklahoma FBI Counterterrorism squad immediately put Case Agents, analysts, and surveillance teams to work. My team was put on standby. First order of business: pull out the "Go To War" kits.

"GO TO WAR"

Black waterproof cases, roughly double the size of large suitcases, were stacked ominously on metal racks on the far wall in the team room. Each case was labeled with the name of the operator to whom it was assigned and the date of its last inspection. The WMD (Weapons of Mass Destruction) kits contained gas mask filters, protective gloves, boots, JSLIST (Joint Service Lightweight Integrated Suit Technology) chemical warfare suits, and a PAPR (Powered Air Purifying Respirator).

Additionally, everyone had an equally large bag in our individual lockers where we kept all of the same contents for routine WMD training exercises. Only the senior, fully qualified operators had a "Go To War" case. It was an odd honor to be among the operators that were trained and equipped to risk their lives in CBRNE (Chemical, Biological, Radioactive, Nuclear, and Explosive) environments.

My kit, as well as Big Toe's, included a radiation detector the size of a brick that we would strap on to our forearm or calf. Its digital display would indicate neutron and gamma radiation on a digital scale. I kept it switched to "vibrate only" mode so that if we came in contact with a radiation source on an op, the device would vibrate against my arm, but not beep, allowing us to maintain stealth on approach. Frosty and two other operators kept wrist-mounted gas meters in their kit that kept track of the oxygen levels in the air. Our respirators could keep out poison, but it couldn't create oxygen where there was none. If either type of our meters went off, it was time to GTFO.

The FBI's WMD program existed long before I joined the FBI in 1999, but it was re-invigorated in the ashes of the largest terrorist attack on U.S. soil. Twelve months after the attacks on 9/11, all FBI SWAT teams had been given rotational orders to travel to a Department of Energy facility in Albuquerque, New Mexico, to participate in tactical WMD training. This was the first training of its type for FBI SWAT, and it was the first of many for me. I would also attend several courses about locating and disarming IEDs (Improvised Explosive Devices) and about nuclear weapons, radiological and biological weapons, and U.S. nuclear weapons transport systems and routes.

We had to still be fully proficient at high-risk arrests, searches, and hostage rescues – but now while sucking air through a rubber mask, stuffed into a charcoal-lined suit, and with a small vacuum cleaner strapped on our backs that kept us alive in a CBRNE environment.

Even after graduating from the three-week SWAT Basic Training at Quantico, it took a lot of additional training for an operator to become WMD-proficient. The senior operators and team leaders had to bring newer operators to full competency in critical skills such as CQB (Close Quarters Battle), medical training, functional fitness, and expert marksmanship while wearing a chem suit and mask, which made everything significantly harder. The communication through an intricate microphone system wired into the masks was prone to failure. The mask stole your peripheral vision and made breathing more labored, even with less exertion. Treating a wound or patching a hole in the protective suit was substantially more complicated – not to mention the suit was unbearably hot in the summer.

Getting operators proficient in all WMD tactical skill sets was almost like training them all over again. WMD training days were hot and miserable, but we knew the mission was critical.

In Albuquerque, Mag and I got paired up for the timed assault course. We crawled through tunnels and ran through a dry riverbed with our full WMD kit on. Instructors would spray the air with pepper spray to check if our masks fitted properly. If you had a leak, you'd know it soon enough. Around each bend, we engaged pop-up targets with live ammo. The clock was ticking and everyone was watching. If you wanted a good time, you had to shoot on the run and balance speed and accuracy. As I raced down the riverbed with Mag, the gas mask made me feel like I was breathing through a straw. I envied Mag's superior cardiovascular ability. I finished the course with a passing score, but I learned quickly that I had to pace myself if I wanted to run, shoot, and not suffocate.

At the culmination of training, our team and two sister teams assaulted a multi-story building with role-players who posed as terrorists and hostages. We had to neutralize the terrorists, rescue the hostages, and disarm the WMD before it dispersed a poisonous cloud over the city. After engaging terrorist role-players with non-lethal munitions in an epic gun fight, rescuing and evacuating hostages, deactivating a biological weapon, providing medical care, and decontaminating ourselves, we packed up and headed home with confidence that we could execute our core missions regardless of the WMD threat. Despite this new-found confidence, I doubted any of us really wanted to experience a WMD operation.

In fact, on the ride home, I looked over at Lil Toe as we led our convoy of SUVs. "Can you imagine having to execute an op with a real WMD threat? That would be a bad day, man."

Lil Toe looked back at me. "If that day ever comes, we should all call in sick."

DON'T EAT THE PIZZA

The Tammy/Tommy and Samantha case wasn't our first WMD rodeo. Three years prior, our first WMD mission had been an attempted homicide case. The subject had made the mistake of admitting to one of his friends that he had figured out how to make ricin and asked for assistance in poisoning his ex-girlfriend, who was pregnant with his baby. He'd told his buddy he didn't want a child and he didn't want to pay child support, so killing the pregnant woman was his solution.

Dangler was the Case Agent on that case. He had contacted the girlfriend and told her about the murderous plot – you can imagine her initial disbelief. Eventually she had agreed to let us take her to a safe location with an FBI protection detail. With us recording the conversation, she had returned her ex-boyfriend's call. As we'd predicted, he had hurriedly asked her to meet him for pizza. The terror had been plain on her face as the truth sank in: he had intended to poison her food.

We'd told her to agree to the meeting, and then we had set surveillance on the subject's house.

When it had come time for the pizza date, the subject had stepped out of his house and been immediately intercepted and detained by plain-clothes full-time surveillance FBI Agents known as SOG (Special Operations Group). He was placed in a local police car until the search warrant on his house could be completed. During that time, he'd tried to ditch a vial of ricin in the back of the police patrol car, but the police had caught it. The fact that he'd put the vial in his pocket when he was headed to see his ex-girlfriend and dumped it when he got put in a squad car, added to his guilt and intent. Long story short: he went to jail, and the lives of the woman and her unborn child were preserved.

The "Pizza Case," as we called it, had gone well, but this case – with a plot to kill a terrorist taxicab driver half a world away – had its own investigative and tactical challenges.

NO CURE

An FBI WMD representative from CIRG (Critical Incident Response Group) had flown in to review our plan and advise us for the "Samantha case." CIRG is the entity in the FBI that coordinates all crisis units: tactical helicopters, surveillance aircraft, HRT, SWAT, ground surveillance, HazMat teams, even a couple of jumbo jets on standby with equipment to disarm a nuclear weapon. All of it was, and is, on 24/7 standby to respond to any critical incident in the U.S. The WMD rep reminded us that, unlike nerve agents and botulinum toxin which disrupt neurotransmission and can cause death in minutes, ricin acts slowly. It stops the synthesis of proteins in cells, killing over hours or days. Inhaling powdered ricin causes bleeding in the lungs and suffocation within about three days.

"Your medics will have atropine auto-injectors, of course," the WMD rep said, gesturing to me and Big Toe. The rep was a burly guy with tattoos on his arms that made me wonder if he had been a sailor before joining the FBI. He continued in a monotone voice, "You can inject them directly through your JSLIST suit, but they aren't going to help. If you're exposed to ricin, there is no antidote, and its effects are irreversible and deadly. You probably won't get exposed, but if you do...you're definitely going to die." *Subtle, this guy is not.* I looked around the room to take in everyone's expression. You could hear a pin drop.

It was a little surreal to put on the WMD kit, or "kit up," for a real mission – again. We suited up and taped off each other's wrists and ankles in the parking garage at the OKC FBI HQ where it was well-lit and we could double-check each other's suits. The PAPR, a battery-powered contraption the size of a small pizza box, was strapped to the backplate of our armor. It had two filters and a hose that connected to our masks and provided constant air pressure. It would protect us if the seals on the gas mask failed and would ensure that clean air blew out of the mask and no poisonous air could seep in. It also provided a welcomed breeze on my face during hot months.

After kitting up, we connected adapters to our gas masks which connected in turn to our radios and headsets. We also attached voice amplifiers to the front of the gas masks. With the cacophony of respirators hissing, the muffling caused by the chem suit hood, and the way the gas mask stifled our voices, voice communication without radios or amplifiers was essentially impossible. Plan A was to communicate with radios. Plan B was to use the voice amplifiers. Plan C was hand-signals. We strapped the respirators on in advance, but we wouldn't don the mask until "jocking up" at the *ad hoc* command post, which was near a police department an hour's drive away.

RALLY POINT

We drove our Bu-Cars (take-home cars) separately to park in a dark, muddy field outside the police station where the Command Post RV had been set up. I parked near our armored vehicle, which was delivered by our Tactical Mobility Agents. We called it the "Boss Hogg" ("Hogg" for short) because "Surplus Department of Energy Armored Nuclear Transport Vehicle" took too long to say. Plus, we had all watched the 1980s TV show, *The Dukes of Hazzard*, as kids, and we all knew that Boss Hogg was fat, slow, and white – just like this truck.

The Hogg was substantially quieter than our old military-surplus MRAP or our up-armored Humvees. It was a plain-looking, big white box truck, but it had heavy armor on all sides, and the front and side ballistic windows were more than an inch thick. The Department of Energy (DoE) had given FBI SWAT dozens of these high-mileage armored trucks when it got new ones. We had trained with the DoE Special Response Force (SRF) agents on several occasions. Their job was to protect convoys carrying nukes while riding in vehicles like these. Our relationship to SRF was that we responded if a convoy got attacked and a nuclear weapon was stolen. In that doomsday scenario, 1,000 FBI SWAT operators would track down the nuke, retrieve it by force, and disarm it on the spot. The FBI nuclear take-back mission was just one part of the SWAT WMD mission that no one really talks about.

There were dozens of cars already parked in the surrounding fields, wedged bumper to bumper. This was a cast of hundreds: FBI HazMat response unit, Army 63rd Civil Response Team, CIRG, Evidence Response Team, local police department, local fire department, a helicopter medevac, negotiators, victim specialists, and a biochemist from FBI headquarters. My team rallied up away from everyone else. We formed a circle behind the Hogg on a semi-dry patch of dirt in the parking area, where Vulcher reviewed the operations plan to serve the high-risk warrant. We would call out, make entry, detain subjects, and safely secure evidence of ricin toxin at Samantha's residence. After Vulcher's review, we split off to make last-chance checks to our protective equipment.

I checked in with Vulcher, then took mental attendance of my guys on Gold team before walking into the small police station. Vulcher had taken over as STL a year ago, and Lil Toe had stepped down to be the Blue Team Leader. Inside the police station, the SAC and both ASACs from FBI OKC were sitting at a folding table in the lobby, sipping coffee. The local police chief, a good friend of mine, was sitting with them. I could tell he wasn't enjoying himself.

I walked into a side room and checked in with the boys in the TOC. I liked to see them face-to-face before they became disembodied voices in my headset. I also found our Electronics Technician, Radio, and asked him if there were any changes to comms. He had set up a mobile repeater at the end of the parking lot and an antenna for the long-range duplex radios that the team leaders used.

"You're going to be pretty far from the mobile repeater," he warned me. "If you start to lose signal, switch off your crypto and it will buy you a little more range."

I walked back out into the muggy air toward our armored vehicles in a sea of police cars, military trucks, and rental cars. It was late June and the cicada bugs and humidity were in full swing. I caught up with Vulcher again. He was going over last-minute plans beside the Hogg.

"What can I do for you, bubba?" I said to Vulcher.

"Comms check in ten. Get the word out, please," Vulcher said in his usual calm, steady voice.

"I'm on it." I answered.

I walked around to each teammate and asked him if he needed anything. Simultaneously, I conducted a silent visual inspection of his gear. I looked for little things. Easy mistakes. A common one was to put on the gas mask first, then connect the hose between the mask and the respirator, and finally sling up your rifle. With peripheral vision limited by the mask, an operator wouldn't be able to see that if his rifle sling rested on top of the respirator hose, it would cinch it off. Since the respirator wasn't running yet, you wouldn't realize your hose was blocked until you turned it on at the last minute, and then it would be too late.

We silently re-formed our instinctive circle in the field outside the police department. Each operator stepped aside to let in the next and the circle grew until it was full, but no one stood inside the circle. Vulcher led us through comms checks using our short-range simplex radios. Starting with OC-1, each operator called out his designator. If you copied your teammate's transmission, you held up a thumb. If you didn't get all thumbs-up, you stepped out of the circle and found Radio or one of the other ETs to do a quick triage of your radio, antenna, wiring harness, microphone, and comms box. The team leaders did an initial check-in with our Tactical Operations Center (TOC) on our long-range radios.

No one had had much sleep. Yesterday we ran live fire drills and dress rehearsals from 8 a.m. to 8 p.m. Today we got up at 3 a.m. to kit up, and now it was 5 a.m. Everyone had their masks on, radios checked, and guns loaded. Ready to move.

Gold team and I mounted up on the Hogg. It was old, but so much roomier than the new Bearcats. I sat on the bench on the right side. K2, my Gold team ATL, and South Beach were across from me. I couldn't see their faces, but they had their designator numbers on their shoulders and a patch with their call sign on the back of their helmets, which was our team's personal tradition. Even without seeing their faces, I could tell who they were by the way they set up their kit. K2 always carried three rifle magazines, bullets facing right, on the front of his armor. Some guys only carried one or two mags. South Beach carried his emergency reload in a hard plastic mag case in the center of his armor. I carried one mag on my chest and put tourniquets in the other two magazine pouches. With one magazine in the rifle and my emergency reload on my left hip, that gave me 90 rounds of rifle ammo, versus K2's 120, but I carried additional medical supplies as the team's lead medic.

SOUND OF SILENCE

Once we were all in the vehicles, Vulcher, Lil Toe, and I conducted secondary comms checks with TOC on our long-range radio. We tried our radio calls in sequence.

"TOC, OC-1, Radio check." Silence.
"TOC, OC-2, Radio check." Silence.
"TOC, OC-3, Radio check." Silence.

I could hear Lil Toe and Vulcher, and they heard me, but that was it. We were only a few minutes away from executing the warrant and the long-range comms had taken a crap. I pulled out my cell phone and dialed Radio's number. It took me three tries because the thick chemical-resistant gloves stole my dexterity. I jammed the phone between my gas mask and my headset.

"Radio, can you hear us on the duplex radios?"

"No. They're down. Just... I'll explain it to you later. I'm working on it."

Later, I would learn Radio was being cryptic because an ASAC had just then ordered him to move the mobile repeater tower closer to the command post to make room for parking spots. In order to move the repeater, Radio would have to power it down, lower the mast, and reverse the process. Despite Radio's argument against it, the ASAC insisted. As a result, we lost our long-range line of communication just as we were passing Phase Line Yellow. It can be tough on field Agents when managers don't have tactical experience.

K2 saw me talking on my cell phone. He shrugged his shoulders as if to ask what was going on. I keyed up my short-range radio.

"The repeater is down. We will have to contact the TOC via cell phone." That wasn't an ideal situation since our call might be for a medevac helicopter. K2 nodded. I keyed up my simplex radio to share the news with Lil Toe and Vulcher.

South Beach looked over at K2 and back at me. The red lights in the back of Hogg flickered off of his gas mask.

"Do you hear Money?" K2 asked South Beach.

South Beach turned his head to K2 and shook it.

So South Beach could hear K2, but not me. All our radios had been working perfectly when we rolled out; now it was falling apart. *Maybe one of our radios dropped crypto?* There wasn't enough time for all of us to manually switch off encryption. It was too late to cancel the mission. Time to bust out Plan C.

I reached up to my mask and flipped on my voice amplifier.

"We're gonna use hand signals once we dismount. Copy?"

Everyone nodded and put a thumb up.

SEARCH WARRANT

From the back of the Hogg, I could see the target house coming into view. Everyone tightened his grip on the handholds overhead as the Hogg rolled to a stop. I pulled the release lever on the heavy, armored rear doors and pushed them open with my boot.

"Gold team dismounting." I said through my voice amplifier as I scanned the dark porch that would be our breach point. I had almost been run over by the Hogg more than once by an overzealous driver wanting to reposition the armored truck. "Put it in park," I said to the driver, just to be safe.

The familiar rubber smell of my gas mask got more pungent as my face started to sweat. For now, our respirators were turned off to decrease our sound signature and to improve verbal communication.

Default and Timmy got out first to establish the shield team. This time Default was on shield, with Timmy as his rifleman. I stepped out behind them and grabbed Timmy's collar to let him know we were not moving yet. We stood in a stack behind the cover of the Hogg's armor while we waited for Blue team to finish surrounding the house and setting the perimeter. South Beach and K2 were my primary breachers, so they stood at the end of the stack, loaded up with breaching gear.

Our surveillance teams confirmed that Samantha and her family were still at the house, so negotiators back at the TOC were calling her cell phone to ask her to come out and bring her children.

Lil Toe's voice came over my headset, "Blue team has set isolation." *Well, at least I can hear Lil Toe.*

Our negotiators called Samantha eight times on her cell phone and got no response. The Hogg's driver turned on the siren and used the public address system to call out the subject, but still no response. It was time to manually breach the door.

Once K2 and South Beach had their breaching gear collected, K2 squeezed my triceps. Then I squeezed Timmy's triceps to let him know we were ready to move. We quietly moved in unison as we approached the front of the structure. As we stacked up by the front door, each operator turned on the respirator of the operator in front of him. I could hear all of the respirators hissing in unison through the amplifier in my headset.

I nodded to South Beach and he called out, "This is the FBI. We have a warrant. Come out with your hands up." The voicemitter on his mask gave his voice a robotic sound. There was no response. He knocked and announced several more times.

We were vulnerable standing on the porch. I keyed up my radio, looked at South Beach and said, "Breach it." I suspected he still couldn't hear me, but at least some of the team would hear and know that the breach was imminent. I didn't see South Beach move after my radio transmission. I would later learn that the connector between my radio and the microphone inside my gas mask was occasionally shorting out due to metal fatigue.

This mission was already going wildly different from how the last WMD mission had gone. After the subject of Pizza Case had been put into custody, our SWAT team had driven up to his house with armored vehicles to look for further evidence of a ricin lab. Since there's no antidote or cure for ricin poisoning, we had decided to have only a small element of operators clear the structure.

Vulcher and I had formed up a shield team; he'd carried the shield, and I had been the rifleman. Wylie had our hand-thrown robot, South Beach had been our breacher, and Lil Toe had been the team leader. Mackey, Twilight, Nestle, K2, and Febreze had stayed outside and held the perimeter while we'd cleared the structure. We'd approached the front door of the house, knocked, and announced that we had a search warrant. After a reasonable wait with no response, Lil Toe had given South Beach the order to breach the door. South Beach had swung a sledgehammer and busted the door open.

Oddly enough, we'd had a malfunction during that mission, too. Wiley had thrown the handheld robot and begun to navigate it through the house, but it had malfunctioned almost as soon as it had hit the ground. We didn't bring the robots this time, though. It was still dark out, and the night-vision variant of the little robots were too slow to clear a room quickly.

Now, I was having to revert to manual hand signals. I waved my hand at South Beach so he would turn his head towards me. I raised my fist above my head and hit my helmet twice to indicate the signal to breach.

South Beach swung his ram: *BAM!* No movement. *BAM!* Nothing. Twice more he swung the ram with no effect. *This must be a barricaded door.*

I called out a failed breach over the radio, hoping that someone on Blue would hear it. And they did. I could hear their voices and hear wood splintering on the far side of the house. Then, K2 gestured to South Beach and pointed to the dent in the door frame. We had chosen not to use lights at the breach point because they backlit our silhouettes and would have made us easy targets for someone shooting from inside. In the darkness and with no peripheral vision, South Beach had been hitting the door frame and not the door. He corrected his aim and cracked the door open on the fifth hit. As the door swung open, I kicked on my weapon-mounted light, stepped into the living room and the team followed me.

We didn't use flashbangs, as planned, for two reasons: there were possibly children present, and volatile WMD weapons might have been there. If there was a chance something in the structure could explode, we didn't want to bring the spark. Our lights darted across the living room, landing on nothing. Empty. We repeated our announcement through our voicemitters: "This is the FBI. We have a warrant. Come out with your hands up." As I scanned the area, the light on my rifle lit up the narrow hallway. I stood and listened for just a second, my right thumb resting on top of my rifle's safety. There was no reason to go running into a gunfight, and there was even less reason to run through a house that might be full of toxins. *Surveillance dudes confirmed that this chick was home. Why is it so quiet?*

Then I saw a woman step into the hallway.

"FBI. Put your hands up." I'm sure the voice amplifier made me sound as much like Darth Vader as the navy blue chem suits and black masks made us look like him.

"Why are you in uniforms?!" she shrieked, absurdly. Then she turned and sprinted down the hallway. She appeared unarmed, but this was a textbook scenario from New Agent Training in Quantico: Always consider the possibility that unarmed people, who are disobeying commands and running away, are doing so because their intent is to arm themselves in the next room.

I keyed up my radio. "Contact! Subject is running toward black side."

She tried to squirt out the back of the house, but Blue team caught her and put her in cuffs. "We got her." Lil Toe called out over the radio. "The rest of the family is in a fifth-wheel in the back."

My team slowly continued to clear the structure, unsure that there weren't other people, booby traps, or biohazards in the house.

Back during the Pizza Case – the one previous time I'd worn the go-to-war chem suit – Wylie and I had been clearing the house together. We'd been in a hallway too narrow for us to stand two-abreast, especially while wearing armor, a chem suit, and a respirator. Our solution had been to simultaneously button-hook into opposing rooms, lean in as far as we could see, then pull back into the hallway to switch out. After we'd both raised the muzzles of our rifles above eye level, we had swung our muzzles down in unison, each turned 90 degrees outward, and button-hooked into our opposing doors simultaneously. It had been an elegant, non-verbal dance – for guys laden with armor, respirators, and assault rifles.

As I had then entered the room on the right, I had seen a room full of cameras, lights, and the umbrella things that photographers use. It was set up as a photo studio – probably for something terrible, in hindsight. But as I had cleared the room and turned left, I saw a closet with the folding accordion doors torn off. Spilling out of it were beakers, coffee filters, a centrifuge, and other laboratory equipment. A shock had gone through me. It's one thing to hear some dude with a Ph.D. in chemistry give a presentation on how ricin is made, and it's another to actually see a clandestine chemical weapons laboratory with your own eyes.

BAD BEANS

Now, three years later, I "stacked" on the entryway to the kitchen and waited. Once I felt a teammate squeeze my arm, I entered the kitchen. I cleared my sector and scanned the room. It appeared as normal and tidy as you'd expect, but something caught my eye. I came closer. On the windowsill by the kitchen sink, there looked to be a dozen black-eyed peas next to a mortar and pestle. No – castor beans. This plant was indigenous to the area, I remembered. Next to them was a small slip of paper with written instructions for extracting ricin toxin from the wild castor plant. No mad scientist laboratory here, but the evidence was just as jarring to see.

The fact that I might be in a house filled with active toxins made my heart rate pick up. I fought the urge to hold my breath. I could feel the respirator blow clean air past my face and out the seals of my mask. *You're good. Steady. Watch for snags. You don't want a hole in this suit.*

We stacked on the next door and cleared the master bedroom. All of the clothing in the house had been put in plastic bags and sealed up. I got a chill up my neck. The bedding was removed and sealed into bags as well. The windows and doorways were sealed with plastic sheets. Maybe she realized that the environment in the house was dangerous for her own family and had evacuated them to the trailer.

The next door in the corridor looked like it exited the structure. "Gold team coming out, black side," I called to let the guys know I was opening a back door. Blue team was still holding the perimeter and Lil Toe was standing next to Samantha in her shiny new bracelets.

"Why is all of your clothing and bedding sealed in bags?" I asked her.

Her face was wadded up with apparent anger. She hesitated. "Bed bugs."

"Really?" I said. "You don't get bed bugs from castor beans."

I keyed up my radio, hoping someone would hear us, "TOC, be advised, the structure is hot. WMD precursors are in the kitchen."

"TOC copies. We will advise ERT and HazMat teams." Thankfully Radio now had the repeater back up and running.

CUSTODY BATTLES AND SICK DAYS

During decon (decontamination), we stripped off our uniforms and wiped down our gear. Because of my last WMD op, I had learned that I'd better wear something under my uniform that looked good on TV. Decon was usually performed outside and the news helicopters were going to show the world your preference in undergarments on a live-feed.

Once I was in clean clothes, I waited for everyone else to clear decon. The HazMat team took over the crime scene and notified us that the house tested positive for ricin toxin. We started to tell jokes to let off steam. I saw South Beach come out of decon and walked up to him.

"So, is that a new breacher technique? Hitting the door frame four times before hitting the door?" I smiled at South Beach. "I'm no breacher, but it seems like hitting the door right away would be better."

South Beach rolled his eyes at me.

I said, "Hey, I'm just fooling with you. Good job on the breach, brother. It's hard to see anything out of those masks, especially with no white light. I was actually thinking about asking Vulcher to buy us some elevated mounts so we can see the rifle optics better when we're masked up."

K2 walked by and overheard us.

"Mounts? Snipers are always talking about mounting...each other," K2 quipped.

"Don't be hateful," I responded, wagging my index finger in the air.

"Snipers like to go out in the woods in pairs. How is that not gay?" K2 said. I was laughing so hard that I couldn't come up with a response. The lack of sleep and the adrenaline crash were catching up to me. I was ready to get back to my truck and head home.

Within hours, the Agents from our Cybercrimes Squad were able to prove that the Craigslist ad came from Samantha's laptop. Analysts dug into her background and discovered her real identity: She was a former Director of Nursing for the Israeli Ministry of Labor and was now married to an American military reservist. She was not an actual Israeli secret agent. But she *was* in a vicious custody dispute with her ex-husband who had recently filed a suit demanding visitation rights. He was the "terrorist" she wanted to kill so she could get full custody of their kids. The paper instructions we'd found in Samantha's kitchen had tested positive for ricin toxin. She would get a 37-year sentence for attempted homicide by means of WMD.

The muddy field where we had parked our cars hours ago was mostly empty now. I hurried to pack my gear in my truck to start my ride home but still took the time to double check that everything was decontaminated. Several teammates were leaning into the trunks of their cars doing the same thing. A lot of people had left the CP area to head home after a long day, but only a few of us had to consider whether or not we might now unintentionally bring home a lethal toxin. I made a mental note to wash everything in my truck the second I got home. Twice. It's crazy how much I still remembered from our initial WMD training school in Albuquerque from fifteen years ago.

Lil Toe was parked next to me. "Hey, Lil Toe," I said as I stood at the rear hatch of my truck. I shook my head. "Dude, there was refined ricin in there."

"I know, Money," he answered.

"That's the second time we've been exposed to ricin toxin. This can't become a trend. I feel like there should be an almost-got-poisoned-to-death bonus for ops like these. You know?"

Lil Toe nodded as he sorted through his gear. I was trying to make him laugh, but it wasn't working so far.

"Hey. We agreed to something fifteen years ago and you let me down. Twice now, actually," I said.

Lil Toe turned his head and looked over at me, confused.

"We agreed we'd call in sick." I said with a deadpan face.

Lil Toe laughed out loud. "Hell, yeah. Next time, we'll call in sick."

Chapter 9
Operation Mischief Mayhem
Mission #98
August 2017

    I flexed my legs to keep the blood flowing. The last thing I needed was my legs falling asleep. Default and I had positioned our generic, white pickup truck at the far end of a mostly empty truck stop. "Dillon's Plumbing" signs were affixed to both sides. From this position, we would be able to see into the windshield of the target vehicle. In a briefing yesterday, the undercover Agent said he would make sure the bomber would park at the edge of the gas station parking lot and in a direction where the bomber would have the best view of the explosion. We had been out here the previous night to conduct surveillance and pick the ideal spot to park our truck.

    It was hard not to fidget in the back of the truck. I was sitting on a suitcase-sized plastic case that I had placed on the driver's side of the back seat area. The back seats were folded up and I had set up a tripod in the empty space where the seats had been, and my legs intertwined with the tripod legs. My precision bolt-action rifle was locked into the tripod and its 24-inch barrel was pointed at the passenger-side rear window, toward where the target vehicle should be later tonight.

The windows were rolled up while we were in the Surveillance phase. When we moved to Overwatch phase, Default would roll down the back right window from his position in the driver's seat, so I could engage threats without having to shoot out the window. We had rehearsed this scenario dozens of times at our monthly sniper training sessions. Through his spotting scope, Default watched the target area with me – 200 yards west of us.

Two hundred yards wasn't exactly a difficult shot. We were shooting at 1,000 yards just three days ago. The Bureau had sent me to a number of advanced shooting schools, and I had gotten very confident making hits at 1,000 yards with this rifle. I had rehearsed for this op with my SR-25 suppressed semi-auto precision rifle, but I ultimately switched back to my bolt action just for the extra accuracy and reliability. The winds were calm, and it was close enough that there was very little windage and elevation correction I needed to make a headshot.

My rifle was loaded with tactical bonded ammunition, excellent for shooting through vehicle windshields with no deflection and with excellent expansion on soft tissue. On one occasion, I assisted with testing this ammunition at the FBI's Ballistics Research Facility (BRF) at Quantico. Every police department and military in the world looked to BRF to test and evaluate ammunition effectiveness in various scenarios. Everything was scientifically done there. Ballistic gel consistency and temperature were strictly controlled so there were minimal variables during testing.

Not only did the round I had selected perform beautifully by not breaking apart when shooting through windshields, it also didn't deflect when it penetrated the glass, so it hit the intended target and minimized risk to innocent bystanders. We referred to it as "barrier blind." It wouldn't expand or break apart when it hit steel, wood, or glass – it would just punch right through. However, once it hit soft tissue, it would expand to over 60-caliber and create catastrophic injuries. It was just the ticket for shooting through a reinforced window at a bank robbery – or in this case – shooting through the windshield and into the face of a terrorist bomber if he threatened the life of our undercover Agent.

BOMB ON WHEELS

My teammate and fellow sniper, Twilight, was one of the Case Agents on this Joint Terrorism Task Force investigation. In FBI parlance, "Case Agent" meant that someone was the primary investigator on a case, not just an Agent assigned to help and run down leads. Twilight was teamed up with Brian, a metropolitan police detective. Both of them shared a quad-cubicle with Vulcher in the JTTF office in the FBI OKC headquarters. Brian was a great guy, a personal friend, and a grandmaster at frying fresh catfish at Vulcher's annual Independence Day cookout.

This investigation was focused on a man in his late twenties who had made plans to detonate a bomb in the same fashion as the 1995 bombing of the Oklahoma City Murrah Building that killed 168 innocent people, including nineteen children. This bomber had emulated the use of an ANFO (Ammonium Nitrate and Fuel Oil) bomb, but was targeting a bank in Oklahoma City, not a federal building. The bomber originally wanted to target the Federal Reserve in Washington D.C. but later decided to bomb a bank in downtown Oklahoma City.

I knew the undercover Agent who was assigned to this case personally. I had met him when he was a new Agent, but he had transformed from the strait-laced, laid-back former attorney I knew to looking like a tatted-up goon. In one of the mission briefings for this case in the SWAT team room, he got an unexpected call. He stood up and said to the caller, "No. Hell, no. You better get yer fucking mind right! Call me back in an hour," in a convincing country drawl. He hung up, smiled, apologized to everyone, and said it was just a "work call" for a different case. Watching him switch gears that fast was impressive, and it gave me even more respect for the pressure on undercover Agents to keep their various personas separate, yet just under the surface and available on demand.

Over the course of several weeks, the bomber had taken the undercover Agent to scout target locations, discuss motivations, and finalize a plan. The bomber decided on a 1,000-pound bomb to be delivered in a stolen white utility van, known as a VBIED (Vehicle Borne Improvised Explosive Device). An FBI Master Bomb Technician and a civil engineer determined that a bomb of this size would have had catastrophic effects on a downtown metropolitan area. The blast pressures would have been sufficient to topple multiple buildings and launch debris over 250 yards, likely killing hundreds of people and causing millions of dollars in damage.

The bomber was fascinated by the movie *Fight Club*, whose storyline used similar explosives, vehicles, and methods to topple the financial system. The movie's primary antagonist was named Tyler Durden. The bomber often quoted the movie and made statements that he wanted to "do Tyler Durden shit." When questioned by the undercover Agent about killing or hurting innocent people, such as bank security, bystanders, or cleaning crews, the bomber again quoted the movie saying, "You want to make an omelet, you gotta break some eggs."

Construction of the VBIED required containers large enough to hold 1,000 pounds of ANFO, along with the requisite components and electronics, and a space where the VBIED could be constructed without fear of discovery. Once the gigantic bomb was assembled, he would load it into the van, drive it through the late-night city streets, and park the van in an alley by the bank. After that, the undercover Agent and the subject would drive a separate vehicle to the truck stop where the bomber would have a view of the explosion after typing a code into his phone that would detonate it.

In order to ensure there could be no chance that the bombing could ever succeed, Twilight and his team covertly entered the bomber's storage facility, set up cameras and recording devices, and substituted the twenty 50-pound bags of fertilizer with perfect replicas that were completely incapable of detonation. Expert seamstresses were hired to sew the fertilizer bags so that the pull-string looked original and untouched.

Additionally, the phone that the bomber put on the explosive device had been manipulated by the FBI so that the detonation signal would be re-routed to an Agent's phone. When the Agent got that signal, we would have the final piece of evidence that the bomber had been fully committed to his murderous plot. If, for whatever reason, the bomber reacted poorly to the failed attempt to detonate the bomb and suspected our undercover Agent of betraying him, Default and I were the first line of defense to protect the undercover Agent from harm.

DEFAULT

"Would you stop breathing so much? You're steaming up the windows," I said to Default.

"It just looks like a couple making out in a parking lot. It's good cover," he answered.

"Well, if we get compromised and we have to put on a show, I want to be the big spoon," I joked.

Default had unintentionally earned his call sign while he was attending FBI Sniper School. The cadre had set up a competition in which pairs of snipers were challenged to hit a water bottle at 100 yards. It was a simple task, except there were two bottles placed next to each other. The sniper on the right had to hit the bottle on the right, and the sniper on the left had to hit the bottle on the left. It was not that much of a challenge, but whoever hit his target first won. Once the race was on, the temptation to let accuracy slide in favor of speed began to rear its ugly head.

Default was up first. His competitor was high-strung and eager to wear the yellow penny of the "top gun." He fired first and struck Default's bottle, eliminating himself from the competition.

The cadre called up another sniper-candidate to square off with Default. The second challenger fell prey to the same temptation and beat Default to the trigger. Default's water bottle exploded from the impact from his competitor's round. This happened again a third time. Default hadn't fired a round yet, but he was winning. Every time he was about to take his shot, his competitor would fire first, but miss. After several rounds, Default was declared the winner. Sometimes he hit his bottle first and sometimes he just waited for the other guy to miss. Later that day, Default was awarded the yellow penny and a couple of jealous sniper-candidates were vociferous about how Default "didn't really win," he just won it by default.

When Default graduated and came home to my team, he gave a briefing on his experiences, as we always did. After he told the story of being "the winner by default," I gave him high praise.

"What were they thinking?" I said. "It doesn't matter who you shoot as long as you shoot first?" I laughed. "Good for you for winning by default. I'd take you over them any day. I'm proud of you. But we are definitely gonna call you 'Default'." I said it with an emphasis on the "De" so it came out like, "DEE-fault." It sounded more like an Okie word. And just like that, Default had earned his call sign.

SURVEILLANCE

My PVS-22 night-vision optic was already mounted on a rail in front of my scope. My eyeball looked through my standard scope, and then through the PVS-22, giving me night vision capability as well as variable magnification. The nice part of this set-up was that I could use my regular scope at night, and I didn't have to worry with different dope or zero on a separate stand-alone night scope. I had a good view of the gas station parking lot where the bomber would park. His spot was lit up from the lighting near the gas pumps. The night scope had decent clarity, but it caused a little loss of sharpness, and everything was in shades of green.

I flipped the quick-detach lever at the base of the night vision optic and pulled it off. I looked through the scope again, unimpeded by the night vision optic, and was pleased to see that there was adequate lighting at the gas station. I decided to leave the night vision optic off so that I could see the full spectrum of colors.

I knew that the undercover Agent was supposed to be sitting in the driver seat when they arrived, and the bomber would be in the passenger seat. From my perspective, the bomber would be on the left. Since I knew that the undercover Agent was wearing a red plaid shirt, for safety's sake I preferred to see the full spectrum of colors.

The Command Post (CP) could hear the conversation between the Agent and the bomber via a hidden microphone installed in the car, but I wouldn't be able to hear it at my location. I'd be focused on just communicating with the CP, the surveillance teams, my spotter, and the assaulters that were standing by to rescue the Agent and arrest the bomber. The undercover Agent had a failsafe duress word he could use to let the CP know if he was in danger, and he had a duress gesture that I would be looking out for. If either I or the CP got the duress signal, we would take action to rescue the undercover Agent immediately.

I took this time to survey the entire area fully. I peered into every car window with my scope turned up to maximum magnification. Briefly, I put my night vision optic back on. With an infrared flood light attached, I could see through the window tint, even in the dark, and check whether anyone was inside the other cars parked in the lot.

I sketched out a diagram, noting the exits from the lot and the hidden location of my teammates who would assault the bomber's vehicle and arrest him. I scanned to see where people tended to walk from the pumps to the cashier. I took note of local police officers who stopped in for coffee. I annotated the distance to every item. If I had to take a shot at a location other than the pre-designated spot, I would already have a reference point to adjust my dope so I could make a first-round hit. I named certain points with brevity codes on the diagram for quick reference. It was faster and easier to say "pimp-mobile" than to say "that purple low-rider Impala with chrome rims."

Once I finished the diagram, I handed it to Default to get his input, so that we would be on the same page. It was critical that we used the same vernacular to ensure clear and efficient communication.

DOPE

Dope is a term that describes the windage and elevation corrections needed to make an impact on a target while taking into account the distance, angle, and atmospheric conditions such as wind velocity and direction, barometric pressure, altitude, temperature, and humidity.

Dope is a word and not an acronym. Its etymology dates back to WWII soldiers describing how they made estimates for distance shooting. Back then it was common vernacular for a number of things, but in particular, for asking, "What's the dope on this thing?"

In order to ensure my dope was correct, I checked the winds on my phone from a local meteorology site. I could have used my handheld Kestrel wind meter, but hanging it out the window would have drawn too much attention. Besides, even with a 10 mile-per-hour wind at 200 yards, I knew that my windage correction wouldn't be more than 0.4 milliradians, and it was a calm night.

Default briefly rolled down the back window while I looked through my Laser Range Finder (LRF). I punched the button on the top of the LRF as I centered the cross-hairs on the pre-selected parking spot. I was expecting it to say 200 yards. The display read 199. *Close enough.* While the windows were down, the cool air fogged up the optics on my rifle scope, so we decided to leave the windows cracked to prevent that from happening again. It was better for us to be shivering in the truck cab than for our scopes to fog up at a critical moment.

Since we had time to burn, I used my cell phone to check our range using mapping software. It indicated 201 yards. Lastly, I decided to "mil range" some known-dimension objects to triple check my range. Residential doors are usually three feet across. Old cars had wheels that were about fifteen inches in diameter. Newer cars might have twenty-two-inch wheels. (If you looked up the make and model of a car, you could get the exact measurement.) Stop signs are usually thirty inches across. I looked for an object with a known measurement. The license plate on a car near the pre-designated spot was perfect. Plates are always 6"x12". I looked through my rifle scope and measured the plate on the car in milliradians. Using the formula T(in) x 27.776/T (mil) = range (yards), I got 203 yards. Everything was in the ballpark.

UNEXPECTED VISITORS

Default and I were in the middle of telling stories and jokes when I caught a slight movement through the back window of our pickup truck. "Default, we've got visitors." A black and white sedan pulled up directly behind us and stopped. We knew this might have been a possibility. We had parked at the far edge of the truck stop to avoid other vehicles, but I'm sure it still looked odd for a plumber's truck to be parked here after midnight. I keyed up my radio.

"CP, this is Sierra One. It looks like we are about to meet the night shift security guards. We need you to backstop this." The Command Post would call pre-established contacts at the security company and at the metropolitan police to let them know they didn't need to worry about the call they were about to get. Something along the lines of, "There are two guys that look like assassins with a high-powered rifle pointed at the gas station across the street. Send backup!" The red and blue lights on their car lit up and two overweight men got out of the vehicle wearing wrinkled security guard uniforms. The one on the right definitely had a pistol on his belt.

"Take your badge off your belt and hold it in your hand," I said to Default as I did the same. I rolled down the back driver's-side window against which I had previously been resting my back. The guard's flashlight pointed directly in my eyes. I knew it would seem unusually dark in the cab because Default and I had meticulously taped off every light in the truck to prevent us being backlit and to optimize our night vision systems.

I used a tone of voice you would use with an old friend. "Hey, fellas. FBI. I know this is going to sound crazy, but this is a high-risk counter-terrorism mission. Feel free to call FBI headquarters or OKC Police dispatch and confirm." He checked our badges and credentials and nodded.

Then he saw my rifle set up behind me, aimed at the gas station's parking lot. His eyes got big.

"Please step out of the vehicle," he said.

"I absolutely would, brother, but if anyone sees us wearing these uniforms, this mission will be compromised. We can't get out of the truck."

He stared at me. In fairness, if I were him, I wouldn't have believed me, either. He nodded again and looked sideways at his partner. "Ok. You guys are good. Whatever it is you're doing, be safe. Sorry to interrupt." And with that, he and the other security guard got in their cruiser, drove back to the well-lit area of the truck stop and called 911 to report their emergency. Luckily by this point, the head of the private security company, 911 dispatch, and metro police were all expecting the call. The guards were thanked for their diligence and asked to kindly stay at least 100 feet away from the snipers in the plumbing truck.

SHOW TIME

Nightstalker was up and circling at high altitude. It was completely invisible to me, but it could see the infra-red chem stick and strobe I had duct-taped to the top of our truck before the op started. I checked my handheld downlink to view his thermal sensors, but the screen flickered blue, and the signal cut out. *Good evening, Murphy's law. Thanks for coming to the party.* Nevertheless, I could hear the crew on the radio. "Target vehicle approaching waypoint Echo now."

That's our cue. 1 a.m. Right on schedule. I chambered a round in my rifle and made sure the safety was on.

Bombers, unlike drug dealers, usually kept to their schedules. Nightstalker had been watching the bomber since he had dropped off his explosives-laden van at the bank. The bank was closed and the bomb was a dud, but in an abundance of caution, we had set up a ruse to make sure the cleaning crew had the night off and that the building was vacant.

The assault team was hidden out of sight but was listening to the audio live-feed from the bomber's truck, watching the video downlink from Nightstalker, and watching the car's tracker location as it drove toward the takedown site.

I scanned the gas station parking lot to look for the bomber's car. It might sound risky to put a sniper overwatch team in a position where we were shooting across a street, but we had done our homework. Our position was elevated enough to ensure that any round I fired would fly well above any vehicle that happened to drive down the road. The target car now parked right where we had expected. From this spot, the bomber planned to have a nice view of the explosion.

My phone vibrated. Twilight sent out a group text to the team. "Be advised: We don't know if the subject is carrying a gun, but we are certain that he is carrying two knives."

A moment later, Lil Toe responded. "K. We got like 37 guns." That made me laugh out loud.

I looked at the bomber though my scope, turned up the magnification to 15x and studied the details in the car. The undercover Agent was on my right, wearing his red plaid shirt; the bomber was on my left. His countenance seemed calm. I didn't see any guns. No one was under any apparent duress. The undercover Agent hadn't used his emergency gesture or code word.

"OC-1, Sierra One. We have eyes on the subject and the UC (Undercover Agent). The UC is in the driver's seat. Please acknowledge."

When they approached the car for the arrest, they needed to know which side the good guy was on.

"Roger, Sierra One," Vulcher replied. "I copy. UC in the driver's seat." After a short conversation with the Agent, the bomber pulled out his phone. The screen lit up his face with hues of blue. His thumbs moved methodically and then he nodded with an air of satisfaction. He stared adamantly into the night. Toward the bank.

Default was carefully scanning the target vehicle with his 40x spotting scope. He had a more powerful optic than I did so he could gather intelligence and spot my shots if I had to fire.

"I think he put in the detonation codes, Money," Default said.

The Command Post had instructions to let us know when the detonation codes went out, but the radio was silent. The bomber scowled and looked hard over his left shoulder at the undercover Agent. The undercover Agent had his hands in the air. *Is he gesturing "I don't know" or... "Don't hurt me?"* I focused on the reticle in my scope and centered the crosshairs on the bomber's nose so the round would punch through his medulla oblongata and cease all motor functions. He wouldn't even be able to finish a trigger squeeze.

Default keyed up his radio. "CP, Sierra Two, can you confirm if the detonation codes were transmitted?" he asked.

"Stand by, Sierra Two. Confirming," a young female voice answered him. I wished now that I could monitor the conversation inside the vehicle. It would have been a lot to manage since I already had comms with my team's assault package, command post, and NightStalker, but interpreting hand gestures and facial expressions was all I had right now, and Default and I were the eyes for the entire operation at this key moment.

The FBI had sent me to schools to learn the details of building IEDs (Improvised Explosive Devices) for the purposes of avoiding and disarming them. We had even detonated small explosive charges with burner phones from remote locations. There's a bit more nuance to "safe" bomb-making than I had expected. For instance, you shouldn't arm a bomb just by dialing the phone number, because a spam call could set it off prematurely. They taught us that more than one bomber in the Middle East had met his end when an unexpected call came in before he could get a safe distance away. It was safer to call the phone and then activate the bomb by punching in a code. But, entering the detonation code didn't always work right. A bomb not cooperating was pretty normal.

The bomber looked back at his phone and his thumbs thumped away at the screen again. He looked intently up and gazed through the windshield into the distance. He was looking for the bank. If he wasn't so distracted, maybe he could have seen a white truck on a hill across the street 200 yards away and noticed the back window rolled down and a rifle pointed at his head.

The Agent was giving the bomber a chance to back out. This had to be the bomber's decision alone. The Agent's shoulders shrugged as he shook his head. His lips moved. It looked like he was saying, "I don't give a shit. Let's just go." The bomber had a look of disappointment. When the time came, the bomber pressed the detonation code into his phone and braced for an impact that never came. During our trial, he would have a hard time explaining why he was – on three occasions – pressing detonation codes into a device that was programmed to blow up the bank, even after his 'accomplice' said they should give up and go home.

The young female voice from the CP crackled on the radio, "All units, the detonation codes have been sent. You are cleared to execute."

Vulcher responded instantly, "OC SWAT: Execute, Execute, Execute."

Several big SUVs roared out of their discreet locations and raced toward the bomber's truck. I maintained sharp focus on the Agent and the bomber. It looked like a fight was brewing. The Agent took off his hat, and the hair on the back of my neck stood up. That was the signal that we needed to extract him to safety. His face was calm and confident, but the hat signal was a clear indication that he needed help right now. He was unarmed and alone in a car with an increasingly irate homicidal criminal.

"OC-1, Sierra One. The UC is signaling for an extract. No weapons seen."

"Copy, Sierra One. Ten seconds out."

My heart rate picked up a little. I'd have to rely heavily on instinct from years of scenario-based training. If I shot someone in the face through a windshield, there would have to be no other safe alternative. But I wouldn't hesitate to take his life if that was the only option to save the undercover Agent.

The bomber swung a fist through the air. I couldn't hear him yelling, but I could see his mouth open and spit flying. *I wonder if he suspects the UC, now that the bomb hasn't gone off.*

The assault package was halfway there. *This is the longest ten seconds ever.*

The assault team had rehearsed this portion of the operation hundreds of times. Pull up, jump out, distract, breach, arrest.

"Five seconds out," Vulcher's voice was steady as always.

Through my right eye, I had a solid shot on the bomber through my scope. Through my left eye, I could see several large SUV's screeching to a stop beside and behind the bomber's truck. Operators jumped off the running boards of two SUVs and my teammates set an L-shaped formation around the target vehicle. Everyone had already rehearsed his specific role repeatedly. Men were pre-designated to open the doors, disconnect seat belts, and pull the bomber and the Agent out of the car.

Timmy was at the front of the formation and swiftly took his position at the front passenger fender with his rifle pointing through the windshield at the bomber. The bomber was still looking down at his phone. With everyone in place, Timmy flicked on his 1,000-lumen weapon light, shined it directly into the bomber's eyes, and yelled, "FBI! Don't move!" The bomber's face was a look of complete surprise and terror. The strobe from the weapon-light reflected off his face and forced his eyes shut.

I clicked my radio transmit button, "Sierra One reports no change. Sierra is cold." I double-checked that my safety was on, because it was "danger-close" for lethal overwatch. Too close to shoot. I pulled the bolt handle on my rifle up as a second measure of safety and dialed back the magnification to get a better field of view. I was a spectator with a balcony seat now. K2's helmet entered the frame of my scope as he lunged into the vehicle. He and South Beach each grabbed one of the bomber's hands to check for detonators. Bombers often make more than one bomb. If he clacked off a chest rig full of explosives, they would all be dead in an instant.

The bomber didn't know what hit him. He was out of the vehicle before his eyes opened from being blinded by Timmy's light. His knees buckled when K2 and South Beach pulled him out of the car. They tried to stand him up, but it was like watching a newborn giraffe. His legs would not support him, so South Beach held him up like a ragdoll as K2 placed him in handcuffs. Meanwhile on the other side of the truck, assaulters pulled the undercover Agent out and cuffed him as planned, to maintain his cover. He was yelling expletives as he was gently being handcuffed and placed in a vehicle separate from the bomber. His gun, badge, and a well-deserved hot cup of coffee were waiting for him in the SUV. *In character until the end. Nicely done.*

I unloaded my rifle and slipped it back into its case. Default started the truck, cranked up the heater, and we started moving to a rally point to meet with the rest of the team. Mission accomplished. No bomb detonation, undercover Agent brought safely home, bomber in custody. No one in the gas station seemed to notice the arrest. It was just a regular old Saturday morning at 2 a.m. A man opened the door of the gas station to let a woman in. The cashier was watching his little TV. No one was any wiser that several months of counterterrorism investigation had just come to a head with the interdiction of a determined bomber.

Chapter 10

Dirty Bomb

The stray campus cats were weaving in and out of busy feet as students and faculty meandered up and down the sidewalk. It was a beautiful fall day on campus. I was rocking khaki shorts and a t-shirt, but certainly not fitting in like I had hoped. I had to purposefully plan my wardrobe or otherwise I'd instinctively wear typical casual SWAT-guy clothes: cargo pants, wrap-around sunglasses, hiking boots, and a hat with Old Glory on it. This assignment was relatively new to our mission set: Tactical Surveillance and Interdiction, or TSI.

It had been twenty minutes since the target had gone into the academic building across the street. Vulcher had gone in to watch for him. I heard Vulcher's calm voice in my earpiece say, "There are four exits that need to be covered." My off-the-shelf earbuds (as opposed to a tactical earpiece) were connected to the handheld radio in the small of my back under my shirt. I bobbed my head to imaginary music, as I listened to my teammate's instructions while we tracked a terrorist target wearing a backpack that could put everyone on this campus in the grave.

I decided to cover the west exit; there was a bus stop at that exit. "Hey guys, I'm gonna catch the bus," I said into my cell phone. I let them know which location I was picking without using tactical jargon since there were people walking next to me. I walked into the bus stop stall and sat on a bench opposite a couple of college kids. They were diligently staring at their phones and took no notice of me.

Meanwhile, one block over, Drillbit was sitting in a plain-looking pickup truck with his tablet in his lap. The tactical tracking program,

ATAK, was running on the screen, so he could see dots that would tell him where we all were, overlaid on a satellite image of the area. Drillbit had volunteered to spend a few months in D.C. on the dignitary protection detail for the US Attorney General earlier in his career. When he got back, he had a surprise poster in his locker from the movie *Drillbit Taylor* – a comedy about a bodyguard who protected children – and this nickname was born.

He could tell you how much $100 million in US currency weighed off the top of his head and how many pallets of cash could fit onto a confiscated Il-76 Russian transport. He left a successful career in the CIA when he got the opportunity to join the FBI. He was an expert at clandestine surveillance, a skilled automotive mechanic, and an incurable foodie. Our shared love for classic muscle cars, fine cuisine, and quality firearms destined us to be great friends. For a time, we called him "Bourdain," "Mr. Slave" (from a South Park cartoon), and "Shark Bait." Drillbit's call signs were as colorful and multifaceted as he was.

Just then, I saw the target walk out of the academic building and head north up the street. He was too far away for me to catch up to him, and I only had a few minutes before he left my line of sight. I discreetly put my finger on my transmit button, put my phone to my ear, and acted surprised. "Hey, buddy! How are you?!" I said, pretending that I just received a call.

Then I keyed up my radio so the whole team could hear. "You know that guy we were just talking about at Tina's house?" I said. Silence. On the other end of the line, the team knew that I had a report.

"Yeah, I know. Right?" I continued. "Anyway, I just saw that dude at Miog Hall. He's headed toward the parking lot. I'll see if I can catch him and let him know he left his jacket at Tina's place."

I made a point not to use call signs or jargon. That was a dead giveaway to eavesdroppers. Conversations had to seem nonchalant to people within earshot but still provide critical information.

Drillbit heard my report and replied, "Money, you're not burned. I'll relocate you," letting me know that my cover was not blown. Drillbit rounded the corner in his truck and stopped at the bus stop long enough for me to jump in. He drove around the block and dropped me off in a position one block away from the target. "Thanks, buddy." I hopped out of the truck and casually started walking.

The team was tight, and you could see it in how efficiently we were working today. We had recently spent a week in Houston supporting police departments decimated by Hurricane Harvey. We slept in repurposed triple-bunk prison trailers parked in the police department parking lot and took showers in a tent behind the building. It wasn't a fun assignment, but we had built *esprit de corps* that was showing now in our teamwork.

As I was walking down the street, I nonchalantly scanned for the subject. Across the street, I saw Gobbler and Uptown wearing yellow pennies and hard hats and carrying clipboards. The best thing about that particular disguise was that carrying a handheld radio looked perfectly normal. I made a point not to look at them. Staring at them could give away my affiliation with them if I was unwittingly burned. They had a solid fixed position, so I would press on in a mobile reconnaissance role.

This target was known to have a radiological weapon dispersion device, also known as a "dirty bomb." The Attorney General of the United States had issued us "Tornado Authority" – to stop him with prejudice. I knew one of our snipers was peering through his glass from the rooftop of a building behind me. If I had to, I could close the distance with the subject and make a specific gesture at his head, and the sniper could open up the target's cranial vault to turn him off like a switch. For the moment, there was a building blocking the sniper's view. I didn't want to take down a bomber by hand, but that was my fallback option. It was an *in-extremis* option for an uncommon, but serious, scenario that fell directly into our team's Counterterrorism mission.

FBI SWAT had the primary mission to terminate terrorists on U.S. soil before they could detonate Weapons of Mass Destruction. "Tier one" military units were capable of training up for the same mission overseas, but according to the *Posse Comitatus* Act, the US military is forbidden from enforcing the law on civilians inside the US borders, and for good reason. Soldiers should only fight enemies overseas, and police should protect citizens inside the homeland. When you blur the lines and let soldiers enforce the law inside the country, you get martial law, abuse of authority and – eventually – dictators and *coup d'états*.

The dirty bomb was a backpack filled with explosives and radioactive material. If it detonated, it would give off a cloud of radioactive dust. It actually wasn't terribly lethal, but detonation on a college campus would be sufficient to terrorize citizens, give the terrorists political influence, and have an enormous economic impact on the area.

Part of my job as a WMD-certified Agent was to conduct liaison with entities that had legal access to radiological sources in my territory: the local hospital had Cobalt-60 isotopes for sterilizing surgical instruments; the petroleum storage business in town used Americium-241 to look for cracks in their oil tanks; a local construction company used Cesium-137 isotopes in Troxler gauges to test concrete hardness. Earlier this month, a Troxler gauge fell off a truck and went missing, along with its radioactive contents. I had an agreement with all of my liaison contacts that they could call me in the middle of the night to help them find any missing radiological sources – no questions asked. I had earned their trust, and they knew that I would help them recover radiological material quietly and immediately, particularly so that the items didn't fall into the wrong hands.

It was yet another example of the importance of the 400 small offices, known as Resident Agencies, which the FBI staffed around the country. Resident Agents can build that level of trust because they live in the communities they serve.

The subject walked north, then turned left and headed west. After another block, he made a left again. Then left again, back east. This was an SDR, or Surveillance Detection Route. Most people don't take three left turns when they could take one right turn. I was familiar with the technique from tracking hostile Foreign Intelligence Services agents (FIS) in the US. It was an effective way for FIS agents to see if we were tailing them, but it worked both ways: it helped the foreign agent identify and shake tails, but it also confirmed our suspicions about their motives. *You know...that I know...that you know I'm following you.*

I increased my stride and tried to close the gap between me and the subject without seeming overly suspicious. Default and I had joined up at the last intersection. I made a point to smile and laugh with Default as if the two of us were out for a stroll. All the while, I resisted the urge to break into a sprint to close the distance. I couldn't risk losing sight of the subject, but I couldn't spook him either. Just then, South Beach passed us on a bicycle. *Where did he get a bike? Nice move, dude.* Innovation is the name of the game in this line of work. I slowed my step and let South Beach take the "eye" (visual contact).

"He's cutting south across the lawn," South Beach whispered through the radio.

We turned south too, but we were a block apart. Walking parallel to him on adjacent roads, we could get a glimpse of him between each building as we matched his pace. We increased our stride to close the gap until we overtook him and were a block ahead of him. Then Default and I made a quick left and walked briskly between two buildings. Once we were far enough ahead, we turned on to a head-on intercept path with the target. The snipers were still blocked, so this would be a hand-to-hand take down.

As we rounded the corner, I saw the target about 150 yards ahead of us. Default and I erupted in fake laughter.

"I know, man. That dude is nuts!"

"I don't know how he's still alive. You know?"

"Did he tell you the story about when he got lost in Amsterdam?" Our voices were loud and playful. Hidden in plain sight.

Default and I widened the distance between ourselves and made a pocket for the target to walk between us. He could have made a hard right or left turn to avoid us, but that would have given him away, and there were other intercept teams in both directions. This had to be done discreetly. If we were playing our cards right, this would be the first time he had seen our faces all day. People don't go out of their way to avoid walking between strangers who are just having a stroll and telling stories like we were.

The second the subject passed between us, Default and I spun around. We each grabbed one of his hands and rolled his wrists into an aikido wrist lock, separating each finger to prevent it from pressing any detonator buttons. The attack on his joints made his back arch and his head rock back as planned.

"FBI. Don't move an inch," I whispered into his ear. Default looked into his backpack and saw wires connected to a detonator. One touch of the button at the end of the wire and the device would make this campus a ghost town for the next six months. The "greater good" was to shut him down using the skills we had just employed. The risk of grappling with him was too great.

Default put his index finger to the base of the bomber's skull to indicate where the muzzle of his pistol would be if this weren't a training exercise. I flashed him a red "kill card" to silently let him know that he had been intercepted and neutralized. As quietly as it began, it ended.

"Whoops, sorry, partner. I wasn't looking where I was going," I said to the terrorist role-player. I gave him a smile. College kids were walking toward us about 200 yards north. Default brushed an imaginary speck off of the subject's shoulder and patted him on the back. The whole incident took just a couple seconds. No one was paying close enough attention to realize FBI SWAT was conducting a counterterrorism training exercise on campus in the middle of the weekday.

Default keyed up his radio. "Tango neutralized."

My earbuds crackled. "End Exercise. Good job, guys. All operators rendezvous in the parking lot."

Training assignment done. Simulated terrorist eliminated on US soil. Time to get dinner with the guys.

Chapter 11
Cartel Safehouse
Mission #106
September 2018

  I was in the front right seat of the "Hogg" as it groaned its way down the back alley. I pulled out my government-issued cell phone and opened ATAK, a proprietary app. The map on screen indicated the location of our sister FBI SWAT team from Omaha with little green dots. They were on the far side of a red box that indicated our objectives. We were the blue dots. In the center of the screen, my personal blue dot was marked *J$*. The app tracked everyone's position, objectives, helicopter landing zones, topography, and more. Omaha was approaching its target, the north house, as we were approaching our target, the south house. We had federal search warrants for both of these adjacent structures in an otherwise upper-middle-class residential area in Wichita, Kansas. The Kansas City FBI SWAT team was hitting a separate structure simultaneously in another part of town.

  The Hogg rolled to a stop far enough away that suspects at the objective wouldn't hear the diesel engine rattling up the road. *Nice neighborhood. I'd love to live on a street like this. Just...with fewer cartel drug dealers.* Lil Toe jumped out and stood at the back corner of the Hogg while breachers from Blue Team stacked up behind him. I hopped out, snapped down my NODs, tucked my rifle into my right shoulder, and joined the stack. Each operator squeezed the triceps of the operator in front of him, until Lil Toe at the head of the stack got "the squeeze." His eyes were intently upon the alleyway ahead of us. Everyone started moving quietly but swiftly toward the back side of our target. The front side of the both structures were being quietly held by the Wichita Police Department, as well as FBI Special Agents from the Oklahoma City FBI Organized Crime Squad, who had driven up the night before.

The autumn air was crisp and dry. I felt the usual surge of feel-good neurochemicals. It didn't feel like an adrenaline rush; it felt like a sense of well-being. *Serotonin, maybe? Where was a neuroscientist when you needed one?* I was close behind Lil Toe. It was a low-illumination night with a waning moon and light overcast. Through my NODs, the light from back porches made the dark green privacy fences to our right cast light-green shadows across the alleyway.

Since I had limited peripheral vision with NODs down, I took tall steps to be careful that I didn't trip over anything in the alley. I clicked the button on the top of my rifle's handguard with my left thumb to activate the infra-red (IR) laser that was attached. I could see the IR laser through the NODs, but it was invisible to the naked eye. The switch on the bottom of my handguard controlled my white light and green laser. If we were compromised or ambushed before our breach point, we would all use white light to fight it out. Shooting with night vision was fun and sexy, but we were cops, not soldiers. If we shoot, we have to make ourselves known and seen, and we have to positively identify our threats.

The breach point was a double-gate on the wooden privacy fence around the backyard. Breachers quietly finished prepping their gear in the dark. The silence was shattered by the growl of battery-powered circular saws as K2 and Grapenuts began to simultaneously cut the hinges of the gate. It wasn't silent, but it was much quieter than the gas-powered saws that we usually used. It allowed us a little bit of stealth and surprise before we reached the main entrance of the trailer in the back of the compound. We had aerial intelligence to confirm that armed guards lived in the trailer and rarely entered the two main structures. As the saws ground angrily at the steel hinges, sparks launched across the alleyway. Vulcher and Timmy eased a light-weight carbon-fiber ladder against the fence and climbed up to get eyes into the backyard while the fence breach was underway. Through my NODs, I could see their IR lasers pointing into the backyard, but the trailer was still hidden behind the fence from where I stood.

With a crash, the gate fell in, and the trailer came into view. *They'll hear that for sure.*

The assaulters had their rifles up and were covering our breachers as they stomped the gate flat to reduce the tripping hazard. The stealthy part was over, so the Hogg chugged up to our breach point on the gate.

"This is the FBI. Come out with your hands in the air. We have a search warrant. *Policía! Manos arriba....*" Our FBI Special Agent crisis negotiator and polygraph operator, Jose, a former U.S. Coast Guard helicopter pilot, got on the PA from the safety of the front seat of the Hogg and started calling out the subjects, both in English and Spanish.

Voodoo's thermal scan overhead indicated our target was unusually hot, which probably meant it had a substantial number of inhabitants. *I hope this isn't another "cartel pinata."*

Three years prior at Operation Celtic Cobra, my team served warrants for fifty-six indictments in Dallas. At one target, Vulcher was carrying a ballistic shield, and I was behind him on my rifle. There were two adjacent entry doors to the same structure. As we stepped onto the porch, we prepared to knock on the left door. In that instant, the right door cracked open an inch and someone peeked out. Vulcher kicked in the right door, but it rebounded back shut. In the millisecond it was open, I caught a glimpse of what appeared to be a refurbished garage. Vulcher kicked the door open again. There were nearly a dozen big Mexicans around a pool table covered in cocaine, beer and tequila bottles, and there were AK-47 rifles leaning against the walls. Before they had a second to react, Vulcher and I charged into the garage and arrested eleven cartel subjects with zip-ties and a false sense of bravado. When the op was over, Vulcher said to me, "That was a cartel pinata!"

I flipped up my NODs and deactivated the IR laser on my rifle's PEQ-15. "Moving," I called out. I took long, steady strides toward the trailer, resisting the urge to run. As expected, we were exposed while we moved to cover. We had to clear the trailer before we went on to the two houses.

Suddenly, the door on the right side of the trailer opened with a creak, and a male subject stood in the doorway. *That's a gun. That's definitely a gun.* I flipped off my rifle's safety and activated the white light and laser on my full-auto short-barrel M4. The green laser danced small circles on his chest. I could see the red dot from my sight and my green laser intersecting at the same point. He was looking out from a dark trailer into a cascade of 1,000-lumen weapon lights. *He doesn't even know where to aim his gun.*

"FBI. Put your hands up or I'll kill you."

I used this phrase when I wanted to convey to subjects that they genuinely were in mortal danger. "Please comply," or "Don't make me shoot," never had the same effect in preventing a shooting and causing surrender. He was focusing across the backyard and was startled to hear my voice so close to him.

He dropped the pistol. *Good decision, sir.*

"Step out of the trailer. Keep your hands up," I said. He walked down the stairs of the trailer, and a woman carrying a baby came out behind him. One operator handcuffed him and escorted him to a police squad car, while another escorted the mother and child to the ballistic protection of the Hogg. An FBI Victim Advocate and an armed FBI medic, who were on stand-by in a sedan one block away, were called to watch over the mother and baby until the operation was completed. I called out, "Need three," to let my teammates know I intended to clear the trailer and I needed three of us to go in.

"Got three," I heard a voice from behind me. Meanwhile, Lil Toe and Blue Team covered the main residence. I stepped over the loaded .45-caliber pistol at the trailer doorway. "Gun in the doorway." I called out. I resisted the urge to pick it up and unload it. *Eyes up. Look for threats.* I knew that a teammate would unload the pistol once we were in the trailer. The three of us quickly cleared the fifth-wheel, and I clicked the top mic button on my comms box to transmit on Fight-Net, "Trailer clear." Only the Oklahoma team could hear, as Omaha was using a separate, encrypted frequency for their Fight-Net. This avoided confusion if things got chaotic.

Our Senior Team Leader, Vulcher, called out, "Clear structures one and two, simo," over the long-range TOC-Net frequency to make sure Omaha heard the commands and would be making a simultaneous approach with us. After Vulcher's command, Omaha FBI SWAT moved toward the north structure while my team simultaneously moved on the south structure. We passed two enormous pit bulls barking hysterically in their cages. In the cages next to them was a row of pigeons. *Are those courier pigeons? Or… do they eat them?*

Drillbit and Default made a ballistic pocket by the back door with their level-four shields while Timmy had his rifle up, scanning for threats. K2 set a hydraulic tool into the door jamb and rammed it in place with a sledge. It looked like giant steel jaws connected by a long cable to an electric drill. Then we stepped about six feet left of the doorway, the length of the tool's extension cord, and crouched behind Default's shield. K2 had preferentially started using a hydraulic breaching tool after Operation Delta Blues when a teammate got shot through the door as he was prepping to breach with a ram. Bad guys tend to aim for the door, so it's best not to stand behind it any longer than necessary.

The hydraulic tool now quietly eased the door's deadbolt out of the frame, and the door silently drifted open on its own. I listened for footsteps or voices. Nothing.

"Prep a bang," I said just loud enough for my teammates to hear.

Twilight, standing next to me, said, "I got one." We stepped quietly to the open door.

I tapped my Fight-Net mic button and calmly called the order, "Bang and hold."

Twilight threw the flashbang into the middle of the room.

BA-DOOM! It detonated and echoed throughout the building. My amplified headset automatically dialed down the volume, so the bang didn't sound that loud. A second later, a similar detonation echoed back from Omaha's objective. Because Voodoo's thermal sensors previously indicated substantial heat, we were suspicious that it was either heavily occupied or heated as a marijuana grow-house. It should be full of people or full of plants. After the bang detonated, I let the room breathe for a moment. There was no rush to go inside. It was just a search, not an arrest warrant or a hostage rescue. Time was on our side.

I grabbed the back of Default's collar so that he knew I was talking to him. I was careful not to touch his arm so that he didn't get an unintentional indication to make entry. I said, "Lip it."

He placed the left edge of his ballistic shield perpendicular against the left edge of the door opening and began to rotate counterclockwise until his shield was parallel with the door frame and blocking it almost completely. From this position, he looked through the clear ballistic glass in the middle of the shield and lit up the room with the battery-powered lights built into the shield.

"Nothing seen," he said.

I called, "Bang, enter, and clear." If anyone was in the house, they knew we were here. A second bang would buy a second of safety to get into the room.

Twilight lobbed a second flashbang into the house. This was to be his last op before he took over as Counterterrorism Squad Supervisor in Oklahoma City. *I think he's planning to beat my record for most-bangs-thrown in a single op.* The flashbang detonated with a fireball and a thunderous echo. Instantly, Default rolled right and protected us from the unknown area in his sector with his shield. Simultaneously, Twilight tucked his rifle into his armpit, stepped left and snapped his rifle up into his shoulder to clear the left sector. I had trained with him enough to know from the way he shifted his weight and placed his steps that he was most likely clearing left, and I had mentally prepared to fill in and clear the middle of the room.

"Clear right."

"Clear left."

"Clear center," I said

I choked on noisome fumes. A gas generator was running at the far side of the room. It must have been running inside the building all night. *That's what was giving the house a positive heat signature.* After we cleared both floors of the structure, we called all-clear on the TOC-Net. *Omaha must have gotten all the good stuff.*

"OM-1, TOC. All clear on Structure Two. Dry hole." The Omaha STL called over TOC-Net to announce they had no subjects and no evidence to collect.

There was no furniture in our building. No drug dealers. No drugs. No flooring, no interior walls. Barren. But from across the street, lighted living rooms and kitchens were visible, with children's toys in the backyards. This structure seemed to belong in this lovely middle-class neighborhood only when viewed from the outside: it was a gutted industrial warehouse on the inside.

I clicked my mic button, "All clear, structure one. Starting secondary searches." I was letting Vulcher know there were no hostiles in the house and that we were going to start backtracking to see if we missed any parts of the house where one might hide. The initial room-clears were usually deliberate but at a snappy pace. The secondary search was much slower and more methodical. On more than one occasion, we found hidden areas in a house on the second pass.

I walked back down to the front room, and Twilight pointed to a hole cut through the floor into the crawlspace. I nodded silently to Twilight. If a subject was in the crawlspace, he could shoot right through the floor and hit us where we have no body armor. I suddenly felt very vulnerable.

"OC-1, can you bring Cletus up here?" I asked.

Now that the primary clears were done, Vulcher's job of quarterbacking the simultaneous entry was over, and he was available to join us in the search.

Lil Toe and Vulcher brought in the Avatar robot. It was about the size of a carry-on suitcase and had rubber tank tracks. Mackey had given it its nickname and it had stuck. Vulcher dropped it in the trap door that led to the crawl space while Lil Toe used the controller with a small screen that showed the robot's cameras. We stood silently as Cletus crawled around under us. I visualized bullets ripping up from the floor and thought about how I would fire right back into those bullet holes if I had to. The camera on the robot did a final 360-degree sweep with its night-vision camera.

"All clear," said Lil Toe.

"Hey, we got something." Twilight's voice echoed through the empty house.

Twilight pointed to a pile of filthy clothes, covered in fecal matter and trash. "There's a hatch under that crap that leads to a basement." Vulcher quietly stepped up behind Twilight, secured his rifle behind his back, and drew his pistol. The pistol was less powerful but much more maneuverable in a tight space. Vulcher squeezed Twilight's left triceps, and Twilight crawled into the hatch, knowing that Vulcher would be immediately behind him. Twilight called out, "Small space," to let me know there wasn't enough room for me, and they were better off clearing it without me. There were a few painful seconds of silence.

Vulcher and Twilight crawled back out of the hatch.

"Nothing," Vulcher said. "It's empty."

We had come all this way, and there was nothing in the house, in the attic, or even under the house. I started walking out the back door when I heard a familiar voice.

"Hey, Money!" An operator was calling to me from the backyard. He looked identical to us except his shoulder patch read "OM" where ours had "OC."

"Febreze!?" I gave him a hug. He had transferred to Omaha Division a few years ago and was now an operator and medic on Omaha's team. I hadn't seen him since the Oregon Militia stand-off on the Malheur Indian reservation two years ago. There he and his team had bunked in tents across from the metal barn where our team slept. It was a small world, and the FBI SWAT community was even smaller. It was crazy how often I ran into old friends around the country.

"Have you seen the hole in the backyard?" he asked me.

I followed him quickly over to it.

Outside the north structure was a jagged hole in the back porch concrete. It was big enough for a man to jump in with his arms outstretched and still not touch the sides. We had gathered some intelligence that tunnels might be dug between the structures to secretly move drugs. Maybe this was an entrance. The Omaha STL slung his rifle behind his back, pulled out his pistol, flipped down his NOD's and said, "Need two." He descended the ladder into the dark hole, and a teammate followed him. I was impressed that their STL went first into the hole without hesitation and his teammate quickly followed his lead.

His radio crackled. "TOC, OM-1. There's a tunnel. It looks like…" His voice garbled in my headset. The concrete was thick enough to block his transmission.

His voice came back for a second. "…between the houses. I think it goes all the way…" and he cut out again.

I saw a light flicker in the hole, and I heard a voice. "Coming out." OM-1's head popped up. His NODs were up and the headlamp on his helmet was on. He was shaking his head as he made eye contact with Vulcher.

"It's crazy, man. They're tunneling all the way between houses. But there's no dope, no guns." It seemed all we had come here to do was arrest two illegal aliens and seize one pistol.

Vulcher pressed his mic button. I heard his voice next to me and heard it transmitted over my headset simultaneously. "TOC, OC-1. Both structures are clear."

Where is everyone? Where are all the guns and drugs?

As the sun was coming up, the Evidence Response Team showed up, and we began to help them search the interior of the house. Some street Agents from Mackey's Organized Crime squad came as well.

Mackey had recently stepped off the team to serve as the Organized Crime Squad Supervisory Special Agent (SSA). We were all sad to see him leave the team, but happy to see a seasoned street Agent like him move up into a management position. Every SWAT team has an SSA assigned to it as a "SWAT Coordinator." They were supervisory liaisons to the team leaders, but they didn't have to have any tactical experience. Sometimes they were just a roadblock to getting the job done. Since they were not operators, they had no authority to give orders or make tactical decisions, but they sure could ask what-if questions *ad nauseum*. We were excited to see an experienced former operator like Mackey take the role of SWAT Coordinator for our team.

I can't believe this is a dry hole. What are we missing? The frustration was palpable.

"Wasn't there supposed to be a cache of weapons here?" I asked Mackey.

"Yup. On the T-3, they said somethin' about hidin' them unna the stairs." A 'T-3' was a Title-III wiretap operation. It took a staggering amount of time, effort, paperwork, and court approvals to get a T-3, but it was arguably the biggest tool in our investigative tool belt.

I looked around the staircase. "There's nothing under the stairs," I said. It was just a freestanding staircase. I stood under the stairs and waved my arms to my sides as if to indicate the lack of drugs and guns under the stairs.

Mackey stood under the staircase with me and ran his hands along the wall. His eyes got big. He turned his palm toward my face. It was covered in wet paint.

Mackey nodded to Red. "Behind the drywall. Whatchu waiting for?"

Without a moment's hesitation, Red smashed a hole in the wall under the stairs. Red was a brand-new FBI Agent and a former US Navy SEAL. He still had to serve his probationary period as a street Agent before he could attempt selection for FBI SWAT, but he had already made his interest known. *I hope Mackey's right. If we jack this house up, let's at least find those guns.*

Red tugged at a sheet of heavy plastic wrap inside the wall between two studs. Mackey tore the dust-covered wrap off an object and pulled out an AK-47. And another one. Then a whole bundle of AKs. Once we had them all out, Lil Toe gave me a look and punched in the next section of wall. Sure enough, there was another tall bundle of guns packed between the studs. This time, a dozen shotguns and MAC-10s. We looked for more patches of fresh paint and punched through more drywall, uncovering dozens of semiautomatic pistols, a Desert Eagle 50-caliber, dozens of AR-15s, silencers, and magnum revolvers.

We took the guns out to the back porch and lined them up for the evidence team. We had discovered a stash of over 300 firearms. Some of them were exotic, and I would have happily added them to my collection, but I knew they would all be destroyed. As our intelligence had led us to believe, the drug dealers traded guns for drugs or vice versa. These guns, as well as the sports cars parked in the back, were currency for the cartel operating in Kansas.

Along with all of the guns was a cardboard box the size of a microwave oven. It struck me as curious. The box was filled to the top with tightly wrapped black bags, and inside the bags was a yellow crystal substance. An Agent from the Evidence Response Team was walking toward me. He was wearing typical blue body armor with yellow FBI lettering on the chest. He collected a tiny bit of the crystal and put it in a test kit the size of a tea bag, shook the test kit, and we watched it turn blue.

"Meth?" I asked him.

"Yeah. That's a shit-ton," he said.

The twenty-five-pound box would have yielded 11,300 one-gram baggies. At $100 per gram street-price, this was over a million dollars' worth of meth. And this was the Mexican stuff. It wasn't like the Nazi-cook clandestine lab dope I used to see in rural Oklahoma trailers a decade ago. This Mexican meth didn't seem to render people toothless, scab-faced, and emaciated. This junk was therefore more coveted, and it probably contained less lye, battery acid, and red phosphorus. It was, ironically, "the good meth." It was a good dope seizure, but it paled in comparison to some of our others. On one mission in Oklahoma City, we seized 1,470 pounds of meth worth $30 million in a single, gutted house the cartel used as a drug manufacturing facility. But getting meth off the streets is a good thing, whether it's one ounce or a house-full.

Also among the guns was a plastic case about the size of a shoebox with a picture of an electric drill on it. It seemed out of place. Twilight was standing next to me on the back porch as I opened it up, fully expecting to find another gun, but saw what looked like a box of charcoal briquettes.

We both suddenly rocked our heads back from the intense fumes. I'd never smelled something that generated genuine fear in me until then. We were overwhelmed with an acidic, vinegar smell. It was alarming and nauseating, and I immediately felt lightheaded. I thought about the Narcan in my medical kit. If I was about to inadvertently overdose on opiates, I'd need Narcan to cancel out the drugs.

I set the box down, and Twilight and I paced a circle around the table with the box on it, taking deep breaths until we recovered. I waved over one of the Agents on the evidence team to test the contents of the plastic case.

"That's black dragon," he said. "Typical Mexican cartel black tar heroin. You won't get a contact high from it. You smoke it or inject it."

The spacious backyard was completely paved with concrete, and there were expensive sports cars and off-road vehicles, perhaps a dozen of each, parked bumper to bumper. "We're gonna seize all of this." Mackey gestured to the cars, drugs, and guns spread across the backyard. In the daylight, we could see a trail of a half-dozen extension cords slithering out of the fifth-wheel trailer and leading to the generator in the south house's living room. *Oh. That's why the generator was inside the house.* The guard who lived in the trailer powered it with a generator, but put the generator in the house to muffle the sound and prevent suspicion.

The brazenness of the Mexican drug cartel always alarmed me. In 2012, our team executed the warrants for Operation Failure to Yield which identified the violent Mexican Los Zetas gang leader and drug lord, Miguel Trevino. His brother, Jose Trevino, bought an enormous horse ranch in Lexington, Oklahoma. There were 400 horses on the ranch, and much of the horse racing revenue and breeding revenue was to launder millions in drug money. Miguel had boasted on an FBI wiretap that he had killed over 300 Americans, and had made mention that he intended to buy a blue Ferrari when he "retired" and moved to Oklahoma. As we searched the ranch, we saw that Jose had horses named "Cartel," "Blues Ferrari," and "My Brother's Secret." He had photos of his family holding up Zetas' gang signs at horse races. Cartels operated with audacity all across our country, and this wouldn't be the last of it.

K2 was walking past me to the Hogg. I said, "Nice job on the breach today, brother. I'm a believer in that hydraulic tool." K2 nodded and replied, "Thanks, man. No need to get shot standing outside the door, right? Nice work on the sniper stuff today." He paused for effect. "Oh. yeah. There wasn't any sniper stuff today. Again. Huh."

I grinned at him. "Hey, K2, have you read all the books about famous breachers?"

"No," he said.

"Yeah. There aren't any." I winked at him.

As I walked away, a middle-aged, well-dressed woman from the house across the street walked over to us. A patrolman on the perimeter had just talked to her, and he gestured a thumbs-up to us to indicate he had checked her for weapons and warrants.

"I live on this street," she said, "and I've known something was wrong for a long time. Thank you for cleaning up our neighborhood."

That comment meant the world to me. There had been so many occasions when we had served warrants in residential neighborhoods where salt-of-the-earth families with little kids lived right next door to a murderer, drug dealer, or child pornography producer. This is what I signed up for. I wanted to be the guy who made life better for upstanding people.

I thought about the toddler who had been living in the trailer behind me. Powered by one extension cord. Surrounded by expensive cars that were traded for drugs. Living in a fifth-wheel next to a house with walls lined with drugs and guns. *What's going to happen to that poor kid?*

I flashed back to two years prior during Operation Dog Pound in Tulsa. We served a high-risk warrant on the house of an armed and dangerous gang member known as Psycho. As soon as we called him out on the PA, a pre-teen girl ran out the front door, yelling, "He's not here!" Her tears made clean streaks on her dirty face. "My little sisters are still inside! HE'S NOT HERE!" She choked on tears drawing in her next breath and waved her hands in the air like she was commanding us to stop. She knew her daddy was a bad man. She didn't even seem surprised we were there for him. She was terrified, yet standing boldly on the front porch to protect her sisters. With a little coaxing, we convinced her to walk up to our armored vehicle. By protocol, we should have used flashbangs and tear gas to draw out the criminal. After that, we would have sent in robots to find and clear threats. *We can't throw flashbangs and gas if there are innocent kids in there.* Vulcher and I made eye contact, and he nodded to me.

"Enter and clear," Vulcher called out. He didn't say, "bang and clear," and he didn't say "mask up and deploy gas." We were on the same page. There was a very real chance Psycho had sent his own daughter out to mislead us. Maybe he knew we would throttle back our tactics to protect the children – so he could ambush us. No bangs, no gas, no drones, no robots. Just South Beach, Dangler and I, manually clearing room by room. Behind each doorway, there could have been a bullet waiting for us.

When we were down to the last closet, I nodded to Dangler. My heart pounded. There was nowhere else for Psycho to hide if he was here. Dangler pulled open the door and there in a pile of dirty clothes were two preschool-aged girls with matted hair and faces wet with tears. The bigger one, maybe 6 years old, had both arms around the 4-year-old – protecting her from the green-eyed camouflaged giants wandering through her house and calling out her father's name.

The little one looked up at me and with a quivering voice cried, "Mommy left us! Mommy left us!" My heart broke. I gently scooped her up, clicked my mic button and said, "OC-2, coming out. Precious cargo." She wrapped her arms so tight around my neck that it almost hurt. She leaned her head into my neck – in the one soft spot she could find between my helmet and my armor. Dangler picked up her sister and followed me. South Beach led the way back out of the house with his rifle at the ready. Vulcher and Mackey were outside waiting for us. They had already called for an FBI Victim Specialist and state DHS. *What kind of monster leaves his own children alone in a closet?*

She wasn't loosening her grip on my neck. I sat down on the back of the MRAP with her and handed her a stuffed animal and a sucker that I kept in my war bag for just such an occasion.

Poor kid. You don't deserve this.

I snapped back from the memory. Another woman walked up to us carrying the one-year-old from the fifth-wheel. She had tears on her cheeks.

"The man you arrested – I am his aunt," she said with an accent. She bounced the little boy in her arms. His feet wrapped around her hips. "I take heem home wit' me." The DHS worker standing behind her nodded to me and confirmed that they had cleared the boy to go home with his great aunt.

"I just want to thank you. Really." She smiled a sheepish smile and wiped her tears with the back of her left hand.

I took off my sunglasses so she could see my eyes. "Of course. You take good care of him." I smiled at her, and she walked away.

I looked at Mackey, Vulcher, and Twilight. "That's what this is all about: 'the least of these'."

Vulcher answered, "Amen."

Chapter 12
Dripping Springs
Mission #108
October 2018

Vulcher cleared his throat loudly and waited until everyone was silent and gave him eye contact. I was always impressed with his patience.

"Long story short: FBI San Antonio is seeking federal charges for an individual believed to be in Oklahoma. The subject, Brian Steward, has made numerous interstate threats including one to kill the Chief of a Texas Police Department with the help of his 25-man militia. Here are some of the statements Steward has made on social media."

Vulcher looked down at an FBI document and began to read out loud:

"At this point I have three pictures of cops I'm going to kill."

"We're going to lose a lot more cops. People, we need more patriots out there killing cops."

"I'm hoping they'll come for me at home while I have all the booby traps set up. I'll get them all."

"You get the point," Vulcher continued. "A criminal database check on Steward revealed a lengthy criminal history, including assault, unlawful possession of firearms, robbery, burglary, possession of illegal drugs, possession of drug paraphernalia, and criminal misconduct. He's listed as an Armed and Dangerous fugitive and has an outstanding warrant for weapons charges. We have positive intel he is located in an extremely rural area of northeast Oklahoma. The Tulsa SSA (Supervisory Special Agent) has already requested Voodoo for air support. This guy has been a fugitive for the last three years, which explains why there has been no trace of him since 2015."

DIRT DIVE

After the briefing, the team headed to an *ad hoc* staging area and began "dirt diving" or mission planning. The term was familiar to those who were freefall-qualified. It originally meant to prepare for a parachute mission on the ground before getting on a plane, but it had morphed into meaning "mission planning." We pulled out whiteboards to sketch plans and reviewed surveillance footage from the target. We broke into teams and planned for contingencies, actions on arrival, and actions on contact. This guy wanted to kill cops, but we weren't going to give him the opportunity.

We divided into our usual teams of Blue and Gold but mixed up personnel a bit. Each team was staffed for mission outcomes, not just roster assignment. My team included some of my usual Gold team guys like Timmy and Default, who were also sniper teammates. We also borrowed Dangler, a fellow sniper and medic, from Blue team. Dangler and I had coordinated with the Tactical Operations Center (TOC) to have a local ambulance on standby and had requested a local medevac helicopter to pre-position at a small agricultural airport near the target location. We didn't expect a gunfight, but if we ended up in one, we needed to have more resources in place than just Dangler and me with our trauma bags.

The plan was for my Gold team to use the armored "Hogg" and block off the driveway as well as the south side of the mobile home. From our location, we would also keep eyes on the back door on the east side of the trailer, in case he tried to squirt out the back. We were below our normal number of operators for this op, so we asked Nestle to drive the Hogg. He had stepped down from the SWAT team to become a full-time Technically-Trained Agent (TTA). His responsibilities as a TTA included installing court-authorized, sophisticated hidden cameras and hidden microphones. He wasn't allowed to do both jobs. It had been no easy decision for him to step off the team, but we all understood and wanted what was best for him. I was happy to have him back, even if it was as a driver for just one op.

Lil Toe would head up Blue team in our brand-new Bearcat armored vehicle with Headlight behind the wheel and K2, Drillbit, and

Needle-D in back. They would block off the front entrance of the house on the west side, and Vulcher would oversee both teams from the Bearcat. The house was surrounded by forest, so it wouldn't be possible to use the breaching attachment on the front of the Bearcat. The plan was to surround and call him out. Patience would be our primary weapon.

It was a busy season for all of us. Some of the guys were taking time to be with family, some were out of state taking tactical courses, some were testifying in court. The team had been in Kansas the month prior taking down a cartel safe house, together with our Omaha and Kansas sister teams. We had then turned around a day later to go back to Kansas to execute the US Attorney General's Protective Detail while he was in Kansas. Those trips were exhausting and it was thankless work.

Personally, I was in a three-Agent office, and both of my colleagues had set their retirement for the same day. The second I got done emceeing their retirement dinner, I inherited all of their old cases. Now, as the only Agent covering six counties, I had formed a small Violent Crime Task Force with the local police department in hopes of keeping up with the tidal wave of work. I procured FBI Task Force credentials for two of their best special projects detectives, who were spectacular. In fact, they increased my already overwhelming workload by generating so many good new cases. I had been further slammed because of a stabbing in Indian Country the week prior, plus I had inherited a multi-million-dollar embezzlement case that was headed to a jury trial soon. It never stopped.

The Assistant U.S. Attorney (AUSA) had been wearing me slick on the embezzlement case. He insisted on holding face-to-face meetings in his office, even though it was an hour drive for me. I was exhausted from the formalities of the federal prosecutive system, and I coveted the familiarity that my friends in the police department had with the county District Attorney. The AUSA regularly commented that it was best if everyone in the meetings would wear business suits and ties to be "professional." I understood his desire to get my take on prosecutive matters, but every hour I spent driving to his office to discuss legal strategies and chat about the pros and cons of the sequence of witness testimony, was an hour that I wasn't conducting interviews, collecting evidence, and solving crimes. I couldn't lose momentum with new cases by overthinking the ones that had been solved. My job was to get to the truth of every matter, prove the facts, and present it to the AUSA. He pressed charges, I was a fact finder.

As much as he was a robotic stick-in-the mud in a trial-prep conference room, he was sort of a badass in the courtroom. If you were a defense attorney and you had to try a case with this guy, he'd crush you with a recitation of case law and federal code.

In addition to upcoming jury trials, I was due for my Bureau physical fitness test, I had EMT refresher training, and I had to conduct firearms qualifications. I thought I was getting by on six hours of sleep and working twelve-hour days, but I was really getting by on caffeine and adrenaline.

My guys from Gold met with me in the hotel parking lot to rehearse the plan and "brief back" – that is, recite back to the team your responsibilities as you understood them. Dangler was giving his brief-back when Vulcher walked over to us. I knew he had to be tired. His briefing with the case Agents had lasted until 11 p.m. last night.

Dangler was finishing up, "I'll be responsible for deploying gas, if necessary. I'll use the Hogg or a ballistic shield for cover. I'll leave my bolt gun in the Hogg where I can retrieve it if I am reassigned to sniper duties. My EMT kit is in the back of the Hogg, hanging on the right side."

Vulcher stood patiently until we were all looking at him. "Case Agent says the subject has an AK-47 and has talked about setting booby traps in the house for us. You ready to roll?"

"We're good to go," I answered, "We just finished brief-backs."

"Good deal. Let's head to the staging area."

We climbed into the Hogg and settled in. Usually, red lights would be on in the cab to keep our night vision fresh, but this was a rare afternoon hit. It was a treat to be able to get loaded and ready without a flashlight. Nestle was in the driver's seat of our armored chariot. He had on ballistic armor despite the vehicle being armored itself. He had been shot before and knew better than any of us the value of Kevlar and steel.

We always designated the last place of cover on a mission as Phase Line Green. In this case, however, we were going to be grinding these noisy armored vehicles down rural county roads and between run-down trailer homes near Dripping Springs, Oklahoma. Once we left the staging area, we weren't stopping. We would keep rolling past Phase Line Green. Hence the term, "Rolling Green." Nestle eased the big diesel giant into position and the dark green Bearcat pulled in behind us with Vulcher, Lil Toe and the rest of Blue team. I pulled back the base of my left glove to see my watch. It blinked from 15:44 to 15:45. It was the hottest part of the day, and we had a fifteen-minute drive between the staging area and the target location.

"Execute, execute, execute." Vulcher's voice calmly called over the TOC-Net radio that only Lil Toe, the TOC, and I could hear. "Rolling Green."

I looked at Nestle and said, "Take us out. Rolling Green."

He shifted the Hogg into drive, and our big white armored box truck belched as it accelerated down the road. The next time it stopped, we would be within stone-throwing distance of an angry man with a rifle who wanted to kill police officers.

SHINY NEW BEARCAT

I felt good. I was well-rested and well-equipped. I had actually gotten a full night's sleep for once. Sleep deprivation was always a concern because of our usual 3 a.m. wake-up and 6 a.m. hits, but since the warrant had not gotten signed until late today, we were executing the operation in the afternoon. There was always a quiet electricity in the air before a high-risk op that kept the adrenaline and dopamine flowing and made me feel sharp and motivated.

I let my carbine hang from its sling around my neck and held it muzzle-down between my legs while I tried to relax and mentally rehearse while the truck was underway. I had been issued many different optics and carbines (that is, short-barreled rifles) over the years. My current set-up was about as much a "do-it-all gun" as I could hope for. With an 11.5" barrel, it was short and handy, but I could still hit targets out to 600 yards with it. That accuracy was largely due to the increased magnification from my ACOG 4x scope.

Additionally, I had an RMR red-dot sight mounted on top of the ACOG that let me shoot quickly during Close Quarters Battle without having to crane my head down onto the stock. It had the added benefit of being high enough over the bore that I could see the red dot when I was wearing a gas mask or night vision goggles.

My rifle was originally fully automatic, but I had been convinced by an FBI armorer at a week-long shooting school at Midsouth to swap out the trigger mechanism with an SSA two-stage match trigger pack. As a sniper, I was picky about the finer attributes of a good trigger: over-travel, creep, reset, stage-weight. This trigger was outstanding. Fully automatic rifles are sexy and everything, but the truth is I had never needed fully automatic fire, plus the match-grade trigger was so good I would never go back.

I had spray-painted a camouflage pattern onto this rifle. I didn't want a rifle giving away my position when I was out in the woods. The plate carrier on my torso was actually pretty comfortable, and my new padded helmet accommodated my headset without pushing it into my brains. I had our custom armored plates inside the front and back of my plate carrier, but I hadn't put on the optional side and shoulder armor. Weight vs. protection was always a tradeoff.

This kit was far superior to the surplus helmets and hand-me-down armor with no plates I had been issued seventeen years ago. Riding in one of our two armored vehicles with air support overhead, our tactics had now become less aggressive – in a good way: we had enough ballistic protection to pause and wait for the bad guys to come out, rather than rushing in – even if they started the gunfight.

Our armored convoy rolled to a polite stop at the entrance to the trailer home community. This was the first operational use of our big green Bearcat. In the last twenty years we had evolved from arriving in old panel vans and Suburbans, to military surplus MRAPs and armored Hummers, to old surplus DOE armored trucks (like the Hogg). Now we finally had a brand-new state-of-the-art armored Bearcat. I was happy that we had gotten the Bearcat and still got to keep the Hogg. This op was a perfect example of why we needed at *least* two armored vehicles.

Nestle leaned out and punched in the code to open the gate. When you have the gate code already, it seems unsportsmanlike to just smash through the gate. It struck me as strange to have a gated community full of rundown trailers. I suppose even the folks in those leaky mobile homes wanted to keep out thieves and criminals. Too bad this gate was protecting an armed and dangerous felon *inside* the park that we were about to have the pleasure of meeting.

I checked our position on my GPS, then I peeked over Nestle's right shoulder and out the windshield. The target trailer was just ahead and on the right.

"Thirty seconds," I called out to the guys in the back of the Hogg. I clicked my radio transmit button and said, "Last turn," before we headed north up the 50-yard-long dirt driveway to block in the subject's truck on the south side. The Hogg creaked to a stop and the air brakes hissed. I held onto Nestle's headrest as the vehicle's massive weight shifted. There were two dogs tied to five-foot ropes on the front porch of the trailer and they were being very vocal about the unfamiliar vehicles in their driveway. *Poor dogs. That's a crappy way to live.*

The Bearcat pulled past us on our left and stopped about thirty-five yards from the front door on the west side of the residence. Normally, we would have paused during the drive to attach a ram to the front of the Bearcat, affectionately and inappropriately known as the

"Dong of Justice." Unfortunately, big elm trees were thick around the target trailer and blocked us from getting within Dong-range. Otherwise, the Bearcat would have driven up close and popped open the trailer's door with the ram. That was always safer than sending out a breacher team to stand on the front porch and swing a ram. If you can open the bad guy's door from the comfort of an air-conditioned armored truck, that's the right decision.

LEAVE A MESSAGE AT THE TONE

The PA blared instructions, "Brian Steward, this is the FBI. We have a warrant to search the premises. Please come out with your hands up." There was no response. Time was on our side, so the negotiator sitting in the Bearcat repeated the instructions into his microphone. In the meantime, Vulcher called the TOC and relayed that we needed the NOC (Negotiator Operations Center) to call the subject's cell phone as an alternative way of letting him know he needed to come out. Until then, we stayed "buttoned up" inside the safety of the armored vehicles. There was no sense in letting him take a shot at us from the concealment of his windows. The PA squealed back to life every few minutes and repeated the commands. *There's no way he doesn't hear us. That PA is LOUD. The whole county can hear that PA. Heck, his neighbors are probably Googling "Brian Steward FBI fugitive" as we sit here.*

Uptown's voice crackled over TOC-Net. He had been a Marine, Federal Air Marshal, and military contractor before joining the FBI. Now he was a Special Agent on the mobile TOC that was currently located a few hundred yards from us. They were close enough to get a good line of communication and far enough to stay safe. (Uptown would later pass selection and become a fellow SWAT operator, so he hadn't earned the name "Uptown" quite yet.) He was in the TOC with David, our lead TOC Agent. They both were friendly, calm voices during a crisis. David always inspired me with his gentle boldness as a Christian. You could see his faith in every way he lived his life. I don't think there was a more trusted Agent on the team.

Uptown transmitted, "The negotiators attempted to make contact with the subject and left a message for him, but uh…they said to let you know that his voicemail message says, 'Leave a message for 'cop killer.'" That made me wince a little. It was certainly "to the point," clearly threatening, and immature all at the same time.

After waiting thirty minutes and after hucking a few flash-bangs onto the porch to let him know we meant business, we coordinated with the TOC and FBI managers to deploy gas into the residence. If he was in there, he wasn't napping, he was prepping for a fight. The last thing we wanted to do was rush into a place that was occupied by an armed man fixated on killing police.

The negotiator's PA instructions were clear, "Mr. Steward, we are not shooting at you. You will hear a popping noise that sounds like gunfire. We are only firing tear gas into the trailer. Please stay clear of the windows."

We gave him another five minutes to come out, in case he didn't want to replace all of his windows. We couldn't have given him a better heads-up as to how this was going to unfold.

PEPPER TIME

"Gold team, dismount and prep to gas," I said.

Dangler slipped his rifle behind his back and picked up a Heckler and Koch 40mm grenade launcher. He pulled the top off of a sealed container that I lovingly referred to as a "bucket-o'-'nades." It was the size of a ten-gallon bucket and was full of OC (oleoresin capsicum) grenades. These "gas" grenades didn't deploy an actual gas the way tear gas does. They were designed to punch through windows or doors and then break apart on a hard surface inside the structure and splash out a mist of OC or "pepper spray." It was harmless, but we all knew first-hand the misery of OC since we had all been sprayed with it while we were in New Agent Training at Quantico. OC spray sucks. A lot. Even though the OC can't kill you, the canisters themselves packed a wallop since they were designed to bust through barriers. That's why we were notifying the self-proclaimed "cop killer" to stay away from the windows while we delivered his spicy warning.

The Hogg was directly behind the south window of the mobile home, and there was a small shed to our right. It was good concealment, but not cover – meaning it would be good to keep us out of view of the subject, but it wouldn't stop any bullets. As my teammates started to pour out of the back of our truck, they took up logical positions. I positioned myself to the right of the shed with Default, who was covering the south side of the residence with the ballistic shield. Timmy and Dangler took up concealed positions that covered the rear, or east, of the trailer in case our subject made a run for it out the back door.

Timmy was a former enlisted Marine and police sniper who had seen combat in Iraq on the Syrian border. He was tall, lean, and muscular. He was assigned to a small Resident Agency in eastern Oklahoma where he diligently worked violent and heinous crimes on Indian land. He had a reputation as an excellent investigator, athlete, and marksman. And he was a great guy. In fact, his call sign on the team was just "Tim." After years of trying to find a random quirk with which to stereotype him, we just gave up. He had been on the team long enough to deserve a custom-embroidered name tape, but we had no clever moniker for him. After conferring with my fellow team leaders, I had sent the whole team the following message:

> Tim: Because your boringly reliable, respectful, unremarkable, normal, healthy, well-mannered, "gray-man" persona has left us devoid of literally even one, single decent nickname after all these years, and due to your flagrant inability to accidentally break or crash even one small thing of value, we dub thee... "Tim."

After that, his ballistic helmet bore the coveted camouflage call sign tape embroidered with just, *Tim*.

BLOOP! Dangler fired a round from his gas-launcher. It kicked like a 12-gauge shotgun and was equally loud, but it had an almost comical tone due to the large caliber of the gas canisters. The round crashed through the window on the south side of the trailer and splattered purple-tinted pepper spray across the wall. I pulled a second round from the bucket-o'-'nades and handed it over Dangler's right shoulder. He popped the expended shell from the launcher, took the fresh round from my hand, and loaded it without taking his eyes off the target.

BLOOP! Sweat was dripping down my forehead and under my sunglasses, which were already fogging up. It was unusually hot for October – 83 degrees and 100% humidity. We would keep pressure on the subject and be patient, but we wouldn't lose the initiative. After all, he had air conditioning, and we didn't.

Several canisters of gas had now gone into the structure, and I was astonished that there had been no response. This was where things usually got tense: the bad guy knows that he has a limited amount of time before he succumbs to the burning sensation in his eyes and throat. If he has any desire to fight, it usually boils to the surface now.

I blinked the sweat out of my eyes and scanned right and left to maintain my situational awareness. Lil Toe and his team were firing OC canisters at the west side of the trailer from the Bearcat. I handed Dangler another round of OC. *That has to be enough gas to fill the entire trailer. I wonder if he has a gas mask.* The gas in and around the trailer was blowing back at us and making my eyes burn.

Unbeknownst to us, the subject was in the bedroom at the far north side of the structure, and he had tacked blankets over the doorway, managing to keep the gas from bothering him too much.

Better safe than sorry. "Mask up," I told my team. "Timmy and Default, stay on gun. Dangler and I will mask up first."

I reached into the Hogg and pulled my mask from a strap above my captain's chair. I already had a fresh filter on it and a fresh battery in the voicemitter. I clicked the voicemitter on. The little microphone in the mask would project my voice through the speaker in front of my mouth. There was no time to connect my radio to the mask, so I put the boom-mic from my headset close to the speaker in case I needed to transmit. I put my helmet back on and pulled my rifle up to my shoulder to cover the trailer's window.

"Swap out," I said. I realized they probably couldn't hear me from just the voicemitter. I clicked the radio transmit button on my body armor. "Gold team, swap out. Dangler and I are covering."

JUST A SCRATCH

Blam! Blam! Blam! Blam! There was a burst of gunfire. It seemed like it was coming from all directions. I had a delicious surge of adrenaline and a renewed realization of my mortality.

"What was that?" Dangler said. There was no muzzle flash from the south window. The gunshots echoed through the woods.

"Hold your position," I told the guys as I sprinted to the left rear bumper of the Hogg. From there, I could see the Bearcat and the front door of the trailer. I snapped my rifle into my left shoulder so that I could lean around the left side of the vehicle and expose as little of my body as possible.

Blam! Blam! Gunfire erupted out of one of the west-side windows of the trailer. There wasn't a muzzle flash, just flying debris. Debris flying *out* of the window. *He's shooting at us.*

Tink, tink. Bullets ricocheted off of the Bearcat's hull. I snapped off my safety.

My teammates in the Bearcat had their rifles up and were returning fire. *CRAK! CRAK! CRAK!* Their rifles echoed through the woods with a sharp report that the incoming fire didn't have. Vulcher fired from the Bearcat into the trailer window where the debris had flown out. I put my red dot on that window. I made a short exhale as I started to press the trigger.

Then silence. No one was shooting. I desperately wanted to shoot into the window. There was a threat in there, and I needed to be in this fight. It was excruciating. *Just pop your head out and try to shoot at me. Please. Pop your head out.*

Nothing came out of the windows. I scanned the area with my left eye without taking my right eye away from my rifle's optic. No one on Blue team was firing anymore. My team was on the right side of the Hogg where I couldn't see them. *Did anyone get hit?*

I called out, "Gold team, all operators okay!?" I almost grabbed my trauma bag from the back of the Hogg as I ripped my gas mask off, sprinting back to the right side of the armored vehicle. To my relief, all three of the guys were safe and sound. *Thank God.* They were still diligently covering the side and rear of the trailer for any new threats. For all we knew, the fight was still on.

At that moment I understood better the stress that some of the operators had felt after our gunfight at Glass Mountain. Although not directly involved in the shooting, they felt distressed about *not* getting to be in the fight.

I double-checked that my rifle was back on safe and took a deep breath. It took every bit of mental control I had not to shoot into the window. To be shot at and not be able to fight back really sucks.

Vulcher called me on his radio to ask for my situation report. "Gold team, SITREP." I took another breath to slow my breathing. Anxiety is contagious and it was my job to sound and act with calmness and professionalism. My voice needed to be steady. I clicked my radio transmit button.

"OC-1, Gold Team. We're solid. No injuries. Are you guys good?"

"Affirmative. No injuries. He just hit the Bearcat," Vulcher said.

It's the Bearcat's first day out and it already took fire. This is why we can't have nice things. In fairness, the bullets barely left a scratch on it. But still, our truck got shot. I shuddered to think how today would have ended had we executed this op without these armored vehicles.

"TOC, OC-1. Shots fired," Vulcher's voice came calmly across the TOC-Net frequency. "No operators injured."

My team had moved to the back of the Hogg. We had less coverage of the trailer from there, but we were ballistically protected from gunfire. I was waiting for the next burst of gunfire, but it didn't come. *I wonder if he's hit.*

David's calm voice from the TOC came through my headset, "OC-1, TOC. Negotiators contacted the shooter via a tablet he had in the trailer. The subject said he's badly injured. He said he unloaded his guns and you could come in and get him. He's in the far north bedroom and he needs medical aid immediately." Vulcher and I made eye contact from across the yard and shook our heads at each other. Chills went up my spine. Nothing had ever felt more like a trap.

I FORGIVE YOU, TOO

"Negative TOC. We're holding isolation. Convey that he needs to come out if he wants medical aid," Vulcher said. After another sixty minutes of negotiation, the subject finally agreed to come out of the east side of the trailer – his back porch. I took the time to put on surgical gloves and then put my Nomex tactical gloves back over them. I didn't want this guy's blood on me. Injured or not, he was getting handcuffed.

I clicked the transmit button on my plate carrier. "OC-1, we could use a hand. Can you plus us up?" I asked Vulcher.

After a pause, Vulcher answered me. "Roger. Drillbit is on the way." Drillbit sprinted across the yard to assist us with a ballistic shield in his hand.

"Gold team, stand by," I said. "The subject told negotiators he's coming out the back door."

Drillbit stood in front of us with his shield up as we assembled our arrest team.

The back door opened, and an overweight, middle-aged man stepped onto the back porch. He was wearing nothing but black gym shorts and had blood on his chest, hands, and feet.

"Put your hands in the air!" I ordered him. He was slow to respond to my commands, leering over his right shoulder at us with a look of resolve. Then he slowly put his hands up.

"Turn around," I told him. I needed to see if there were any weapons behind his back. Again, he moved slowly. Almost defiantly.

"Walk down the steps and face away from me." I instructed. He didn't comply. He looked down and saw the green laser from my rifle dancing little circles on his bloody chest.

"Do it now!" I said with the tone of a disappointed father. He still didn't comply. Facing me, he dramatically looked up into the sky and blinked. He slowly outstretched his hands to his sides and said, "I forgive you."

"I'm not going to shoot you, you idiot. Just come down the stairs," I said. He had already done a slow 360-degree spin, and it was obvious that there were no weapons on him or his snug, black athletic shorts. He looked at me, slumped his shoulders and faced away as I had instructed. Once he walked back to us, I handcuffed him and started escorting him to the back of the Hogg.

He spontaneously said, "I saw the whole thing with LaVoy and figured y'all would kill me." I could only assume he was referring to Robert LaVoy Finicum who was killed in Oregon by State Police after reaching for a gun and making comments that they would have to kill him. It was surprising to hear him make a reference to the standoff at the Malheur National Wildlife Refuge in Oregon while we were standing in the woods in rural Oklahoma. Little did he know that my team had been there for three weeks during the resolution of that operation.

"No one is going to get killed today," I said.

"I'm glad y'all are the good cops," he said. He paused. I think he was just starting to feel the pain from his injuries. This 260-pound man looked right at me and said, "Can you carry me? The gravel hurts my feet."

"No, I can't carry you. Let's just get behind this truck right now." *I want to get away from the window. I don't want to get shot at any more.*

"You got any friends with guns in the trailer?" I asked him. He didn't answer. Drillbit was walking backwards with his shield between me and the trailer to cover us as best as he could.

I got the subject seated on a step behind the Hogg and quickly went to work to stop his bleeding. When I had treated Nestle's gunshot wound years prior, I had felt like I was moving in slow-motion, like my body was submerged in jello. This time, it felt completely normal.

While patching him up, my mind wandered to some of my EMT (Emergency Medical Technician) training:

HEARTBEAT

Eight years ago.

I'm running over the twenty-foot-high dirt berm to get to my patient. My ears are ringing from the blanks the instructors are firing. I can feel my pulse pounding in my neck. An instructor yells at me: "Get off the 'X'! Move!" The instructor points his pistol in the air and lets out another burst of blanks. "Move to cover!" I scoop up my mangled patent and run back over the berm to a safer location.

Her heart is beating so hard. *She has a strong pulse.*

Her leg is missing and there is a horrible wound to her face. Part of her jaw is gone and she gurgles blood. *Ok. She's breathing.*

Blood is pulsing out of her stump with each heartbeat. *Bright red blood. That's arterial bleeding.*

I remember the photos they showed us in the classroom, depicting soldiers wounded from roadside bombs. I'm far enough away from the gunfire to stop for just a second. I can't delay treatment much longer. *I have to stop the bleeding. She only has so much blood.* When half of your blood volume leaks out, you die. I could start an IV, but that wasn't much use since I didn't carry refrigerated whole blood with me.

I gingerly set her down and instinctively scan 360-degrees for threats. I place a tourniquet over her stump, strap it tight, and turn the windlass until the bleeding stops. The blood is hot and sticky. I can feel the warmth through my gloves. *This isn't time for definitive care. This is time to treat critical wounds that will kill her in just minutes.* She can't keep losing blood at this pace. The tourniquet slides off her stump and her leg starts pouring out blood again. *What the hell?* The tourniquet is soaked in coagulating blood and the Velcro won't stick. I had never put a tourniquet on a real severed leg before. I pull the tourniquet off, run the loose end of it through both loops of the buckle and get enough tension to stop the bleeding even without the Velcro. *That's a start.* I doubt that she is getting good oxygenation by the way she is gurgling, but I have to stop the bleeding first. *If there's no blood to oxygenate, there's no sense in breathing.*

I can see my colleagues waiting at the trauma tent just a couple hundred yards away. *I have to get to them.* It feels like I've been downrange for hours, but it has probably been just a few minutes. I pick my patient back up and run. Inside the tent are a medical doctor, a nurse, and a paramedic. I put the patient on the stretcher and brief them on her injuries. "Catastrophic facial trauma. TQ on amputated leg. Dyspnea. Tachycardia."

The doctor looks at me and says, "Bag her."

I grab a non-rebreather Bag-Valve Mask (BVM), place it over her broken jaw and squeeze the bag to push oxygen into her lungs.

A teammate checks her pulse. "Tachycardic. Thready pulse."

I try to squeeze the bag again, but air won't go into her lungs. *Man, her face is all jacked up.* I try to intubate, but it fails. Her injury looks like a gunshot wound to the face. The patient, alive but sedated, is bleeding out her mouth and choking on it. I can't get air in her lungs from intubation, so it's time to try something different. "Intubation failed," I say to no one in particular.

"Crike her," the doctor says. I had never done that before. I'm not really trained for a cricothyroidotomy, although I have practiced it on human cadavers.

"Me?" I hesitate.

"Do it," he says. I grab a field scalpel, cut a small opening in her neck, pull out her windpipe, nick it open, and put a breathing tube in it. The whole time I feel like I'm hurting her, and I should stop, but she doesn't feel a thing. And she will die if she doesn't get some air soon. I connect that tube to the BVM and squeeze the bag gently. There was no resistance. A full breath into her lungs. *Oh, thank God. Okay. She can get air. This is good.*

The instructor barks at me, "You just gonna let her bleed like that?" I'm not sure what to do. The instructor seems genuinely agitated. "Is she breathing out of her mouth?"

"No, she's criked."

"So?" he says. "Why are you letting her bleed out her mouth?"

I'm confused for a second. I could pack the wound in her throat to stop the bleeding, but I don't want to choke her. Except she's breathing through her neck now, not her mouth.

"Pack her mouth. Never let a patient bleed," he says. I open a hemostatic gauze package and hesitantly pack it into her mouth and throat. All bleeding is bad bleeding. She needs every drop she can keep. I see the instructor nodding at me in my peripheral vision. I resist a fleeting urge to stop everything and soothe her as she dies. But I won't let her die. *These wounds are horrible.* Her vitals start to drop. Her trachea is deviated to one side.

"Pneumothorax?" I look up at the instructor.

He nods.

"Needle decompression?" It feels like I need to ask permission before I jab a needle into her chest cavity.

The instructor gives me a poker face. I have to decide this on my own.

I feel between two ribs and stick a large gauge needle into her chest. There's no hiss as air escapes the chest cavity, but it seems to help a little so I go back to ventilating her.

The doctor on my team says, "I can do a chest tube."

I check for her pulse. I can't find it. "Doc, there's no pulse." I check a second location for a pulse.

"Just a minute," he says. He's cleaning the incision site where he's about to insert a tube that will be more effective than the needle decompression I just did.

I repeat, "Doc. Her heart isn't beating."

"Just wait a goddamn minute!" The doctor is red faced. He's as stressed as I am. "I'm almost done with the chest tube!"

The instructor leans into the doctor's face. "Why the hell are you worried about a chest tube when your patient's heart...isn't... BEATING?" The doctor snaps out of it. "Start chest compressions," he tells the nurse on our team. I keep squeezing the bag that is pushing air into her lungs. I'm looking right at the patient's face and I am overwhelmed with the idea that she is alive right now but may not be in a minute if we can't keep it together. This sudden surge of empathy doesn't seem to be helping at the moment.

"You're doing alright. Okay? I got you," I say awkwardly out loud. I have no idea if anyone hears me, but I suddenly hope they haven't. I don't know it yet, but a few years from now, I'll say the exact same words to my teammate, Nestle, while stuffing gauze into his gunshot wound. Hoping he doesn't die. Hoping I don't let him. Hoping I don't fuck this up and fail everyone.

But this patient's heart isn't beating. CPR isn't working. It's time for more radical efforts. The doctor uses a scalpel and makes a long incision down the patient's left side. He looks at me, "Start cardiac massage." I'm astonished that he's telling me to do it. If her heart won't beat on its own, we have to make it beat for her. This is the last resort. I put my hands together, palms touching – like I'm about to say a prayer. I insert the fingers of both hands into the incision between the patient's ribs and draw my palms apart while keeping my fingertips together. The rib bones crackle as I make enough room to push my gloved hand inside her chest cavity. To my surprise, I find her heart right away. *She's dying.* It seems almost hopeless.

The instructor yells again. "Don't let her die! Come on!" He adds more pressure to the already surreal moment. Her heart feels so small. Warm. I squeeze it. *Am I doing it fast enough?* She has a faster resting heart rate than a human.

"Do something!" the instructor yells at me. "What if this were your teammate?"

I am doing something. I'm literally making her heart beat with my hand. A veterinarian who has been intently watching over this training from the beginning gives the patient a little more ketamine. She feels no pain at all. As trained, I squeeze in rhythm to the Bee Gees' song that ironically has the right tempo. *Stayin' alive. Stayin' alive. Ah. Ah. Ah. Ah...* If I squeeze too hard, I'll crush the cardiac tissue. If I don't squeeze enough, the chambers won't fill.

"I got a pulse!" The nurse says. I stopped the cardiac massage and feel for the pulse myself. It doesn't come.

"I lost it," she says.

I squeeze again. "That's me," I say to her.

"What?"

"You're feeling a pulse when I massage the heart."

"Oh. Yeah," she says.

The patient's jaw is mangled. Crimson-soaked gauze hangs out of her mouth. Her throat cut by my scalpel. Her heart in my hand. *She's dead.* The instructors know that already. It still seems surprising to me. And awful. I feel like a failure.

"End Exercise," the instructor says calmly and makes notes on his clipboard. *So that's it.* I choke back awkward feelings of sadness and exhaustion. *It's okay. It's not a real person. It's fine.* The veterinarian in a white lab coat gives the patient an injection from the red syringe from his shirt pocket. A final euthanizing shot to make sure she doesn't suffer. Blood all over my hands, shirt, pants, some on my face. Bloody gauze and wrappers all over the ground. Time to clean up. We put the patient in a body bag and place her gently on top of the last patient from another team in the tent next to us. We restock med supplies. The next exercise starts in fifteen minutes. *This is going to be a long day.*

When I get home that night, I change clothes in the laundry room and start the laundry. I have the oily old blood from a half dozen Caprine Human Tissue Simulators on me. I take a shower while my wife and daughter nap quietly. When they wake up, my wife asks me what I did today. I say, "medical training."

NEVER SAY "GOAT"

I had assisted in performing autopsies as a Special Agent in OSI, and I had participated in human cadaver labs as an FBI EMT. This exercise had been something different. What is a Caprine Human Tissue Simulator? Well, I know from training that you're only supposed to refer to them as "patients." They give their lives so that we can make stupid mistakes trying to treat them and not make those same mistakes when humans get shot or stabbed or blown up. I know you definitely never call them "goats," and you never disrespect the patients. A US Air Force Pararescueman got dismissed from the last class because he lost his demeanor under stress and failed to treat the patient with utmost respect. That earned him immediate disenrollment.

I'm confident that this training saved Nestle's life the day he was shot. As horrible as it sounds, you can't simulate the stress of keeping a patient alive until you actually have to keep a patient alive. You can't understand the stress of a tourniquet slipping off of a severed limb until you experience it. You don't really know the frustrating effects of coagulated blood on Velcro until you try it. You can read about cricothyroidotomy and cardiac massage and chest decompressions, but there's nothing close to doing it while you're worried that your patient might die, and an instructor is yelling in the background. We cut throats to establish airways. We cracked ribs to pump a heart by hand. We punched needles into the chest of our patients to open up collapsed lungs. It was real, excellent, and awful, and it prepared me for the real thing. I am eternally grateful for the training. I'm grateful for the patients who taught me how to save human lives.

BULLET HOLES

I brought my mind back to the present. I was treating the man who had just tried to kill us. I wasn't the least bit nervous as I treated his gunshot wounds. I had treated many – both on human patents and on other "patients." *Two holes in. One hole out. He's probably got one still in him.*

If people have serious airway injuries, it's difficult for them to speak, so while I was looking at his back for wounds, I was asking random questions to keep him talking and distracted. I could determine his respiratory rate by the way he spoke. In addition to the holes in his chest, abdomen, and back, there were two wounds around his knee where blood was pouring out. Muffled ambulance sirens howled in the distance. *Airway - good. Breathing - good. Circulation - well, we'll see. Keep him alive until they get here.* I quickly put a tourniquet on his leg and tightened it until the bleeding stopped. I used a gauze patch to put direct pressure on his chest to get the holes in his chest to stop oozing dark red blood long enough so I could put an occlusive dressing on it.

A fellow medic, Needle-D, had run over from the Bearcat and was helping me get the dressings out of the wrapping. Needle-D was a new member of the team, but he was a fellow registered Emergency Medical Technician (EMT). Needle-D had gotten his call sign after an extended medical training scenario which I had proctored for the team. In his part of the scenario, the operator he was treating had symptoms of pneumothorax – a condition brought on from a gunshot wound in the chest. In order to treat him properly, the medic needs to relieve the air pressure that is collapsing the lung. Rather than use the term "needle decompression" or even the slang "chest dart," he'd repeatedly used his own abbreviated term for the procedure: "Needle-D." It was not a medical term that I had ever heard used in my twenty years of being a Nationally Registered EMT. He did the scenario well, but after using the odd term on no fewer than four occasions, he had my attention. I complimented him on how he treated the patient, then told him that his name would now and forever be Needle-D. It stuck like glue.

Now I told Needle-D to pull occlusive dressings from my trauma bag while I used direct pressure and gauze on the subject's back to stop the bleeding long enough to get the dressing on. The blood oozing out of his wound kept the dressing from adhering. *Keep him talking. If he's talking, he's breathing.*

"What sort of AK-47 do you have?" I asked.

"Oh, uh, I don't have an AK-47. I shot at you guys with a Highpoint 45."

"Uh huh. Ok." I mumbled, distracted. *The dressing looks good. It should keep air from getting into his chest cavity. Keep looking for injuries.*

"Thank you for not killing me. Thank you for not letting me die." The subject said.

Here I am patching up bullet holes in a man who threatened to kill police, shot at us, and now he's thanking me. Crazy.

"Yeah. Of course. Looks like your ride is here," I said.

"I was usin' Full Metal Jacket ammunition. I knew it wouldn't go through y'all's vests," he added. There was no factual basis for that assumption on his part, but maybe it was supposed to be some sort of indirect apology.

Wasn't going to help him get out of his 32-year federal sentence though.

The gravel crackled as the local standby-by ambulance rolled up behind me. The TOC had launched it without me needing to ask. It parked behind ballistic protection of the Hogg. I transferred the subject onto the ambulance gurney and rechecked his wounds before they transported him. I glanced up at the rural EMTs.

"Hey, guys. Let's get high-flow O2 at 15-liters-per-minute on a non-rebreather," I said politely. *Recheck wounds. Recheck for shock.* I felt underneath the subject's extremities and looked for fresh blood. A local Agent, Dean, from the Tulsa RA was assigned to stay with our would-be-killer en route to the hospital. The subject was wounded and had a right to life-saving care, but he was also under arrest and required an FBI escort. Dean, a former 2nd Battalion Ranger officer, was the Agent assigned to prisoner escort. I knew Dean's name only because I saw it on his body armor. Dean would later try out for SWAT and become a leader on our team. He saw that I had switched fully into "medic mode" and recognized that it was time for me to transfer care to the ambulance crew.

He put his hand on my shoulder. "Sir, he's not your patient anymore. You need to let the ambulance transport him."

I nodded, but I wasn't listening well. "Yeah. Um. Just a sec." I shook my head a little and looked at the ambulance crew. They were staring intently at me. This was no ordinary day for them. A man in camouflage uniform with a rifle on his back was treating gunshot wounds on a shirtless man in the middle of a dirt driveway. No wonder they were staring at me.

I looked up and made eye contact with them, "He's all yours, boys. Multiple GSWs. Good capillary refill. He's stable."

With that, they pushed him into the ambulance with Dean, shut the doors, and raced off. While I had been treating the man who had shot at us, Lil Toe led the rest of the team to clear the trailer and ensure that it was empty. To their surprise, the subject had made snow angels on the floor with his own blood before he finally decided to surrender.

AFTERMATH

Once we were certain there were no other subjects, victims, or wounded operators, the helicopter we had had on standby was cleared to transport the subject to an adjacent state for immediate surgery. With the ambulance gone and everyone still, it was an eerie silence. There was a pile of wrappers from my medical supplies at my feet, and my emergency trauma kit had most of its pouches opened. As I started to piece it all back together, I realized that I had blood up to my elbows on both arms. I peeled off my uniform top and started to wipe myself down with a fresh gauze pack. I asked a teammate to pour a water bottle over my arms while I scrubbed off the blood. *Great. Now I'm gonna have to get checked for hepatitis. Or AIDS. What a crappy way to get AIDS.*

A radio call informed us that supervisors back at the Tactical Operations Center wanted us to stay in our positions until the Evidence Response Team (ERT) could drive out and do an analysis of the shooting. It took hours for the evidence team to get there. We were exhausted and thirsty from the fight. As the sun set, the temperature dropped into the 60's. In the shadows of the tall trees and soaked in my own sweat – and someone else's blood – I started to shiver. It is an understatement to say that we were disappointed when we learned that the command post had ordered pizza while we were shivering in wet clothes and getting bitten by mosquitoes out in the woods.

As the ERT members started to trickle in, one of them asked us to gather up. He was from a small Resident Agency where he worked as a financial analyst. About half of ERT was comprised of Special Agents, and the other half was Professional Support, like communications center dispatchers, evidence technicians, and administrative assistants.

"Hey, I need y'all to line up so we can take photos of everyone. Make sure that you have on your body armor, helmets and uniforms in the exact same way as it was during the shooting," the ERT leader said.

I answered, "I can't. My uniform has someone else's blood on it."

"If you were wearing it during the shooting, you'll need to put it back on," he said.

"It's a blood-borne pathogen risk. I'm not putting it back on." I told him.

"Not my rules, man. Just put it on real quick," he repeated without making direct eye contact.

"No. That's stupid. Just document that I already took it off." I was tired and felt a surge of anger.

"Ok. I'll just call your ASAC and get him involved. You can wait over there while we release everyone else from the crime scene." I knew my ASAC (Assistant Special Agent in Charge) wouldn't back me on this.

I was furious. It all felt very humiliating. Uncharacteristically for me, perhaps because I was tired and hungry and cold, I begrudgingly agreed to put the blood-caked uniform back on long enough for him to take photos. After more random questions from the Evidence Response Team, we all loaded into the Hogg to drive back to where all our individual cars were parked. The Bearcat with its fresh bullet marks had to stay there for ERT to finish photographing. As we were finally pulling away, that same evidence team leader climbed onto the drivers-side running board and banged on the side of the Hogg.

"Hey, can you guys try and use a little less tear gas next time? We have to process the scene once you guys get done wrecking the place," he said. He had a little tomato sauce stain on his shirt. *I wonder if they ate pizza for dinner back at the TOC. Pizza sounds amazing.* I would have happily eaten the cold crusts at this point – if they had bothered to just bring us their leftovers.

We all stared silently at him. Dumbstruck by his question.

"Also, you guys need to get out and stay here for a while longer in case we have any more questions while we're processing the scene." He didn't know when to stop.

"No. We're leaving. Call the SWAT Coordinator if you have more questions," I said. I knew Mackey would tear him up.

"I know you don't like it, but it's easier to have you all stay close in case we have more questions," he said. I was fuming. We had just been shot at, we had been on scene for hours, and none of us had eaten since before we first loaded up in the trucks.

Vulcher gave him a glare and calmly said, "Get off my truck." The ERT technician shook his head and jumped off the Hogg.

I made the two-hour drive home that night sort of in a trance. When I got home, I put my bloody clothes in the washing machine and started it. I tiptoed to the bathroom, trying not to wake up my family and took a long shower. I crawled into bed about 2 a.m., but I laid there wide awake. I just kept rehearsing the gunfight in my head – analyzing every action and second-guessing every decision – until I drifted off to sleep around 3 a.m.

I lunged awake and sat up in bed. It was 6 a.m. I was exhausted but hyperalert. *At least I slept for a few hours.* My mind was in overdrive, so I might as well let my body follow. I needed to go to the office and fill out the medical forms reporting that I had been exposed to someone's blood, and I had to meet with the inspectors that were flying in for the shooting investigation.

MAGIC BULLET

The shooting investigation team was very classy. That had been my experience with other shooting boards, as well. All of the gloom and doom I had heard about the review boards from Agent-involved shooting inspectors had never been the case for me, but then again, we had only been involved in clean shootings. After my interview, I got together with teammates to grab lunch and decompress from the stress of the previous few days. We took up a row of booths at a local burger joint.

I turned to Vulcher in the booth behind me and asked, "Hey, brother. Do you know if that dude survived? He was stable when I handed him off to the ambulance team. I just kinda want to know if he lived."

Vulcher answered, "Yeah, get this." He scanned the room and lowered his voice. "They rushed him to surgery as soon as he got off the helicopter and treated him for the gunshot wounds. But since one of the rounds didn't exit, they opened him up pretty big so that they could get a look inside. They saw that he had cancer on one of his kidneys, so they removed the tumor while he was in surgery. He had no idea that he had kidney cancer until he came out of surgery later that day."

I took a big bite of my burger and pondered that for a little bit.

"So, he's cancer-free?" I asked.

"Cancer-free." Vulcher answered without looking up from his plate.

I paused and let that soak in. Then I smiled at the irony and said, "So, he's lucky that you shot him, right? I mean, you're the reason he's not going to die from cancer!"

Vulcher laughed, "Well, if we cross paths again, I'll tell him he's welcome."

Chapter 13

Operation Vatos Locos

Mission #114
December 2019

The Bearcat's engine chugged lightly, and the airbrakes hissed as we backed out of the garage at HQ. This was a local arrest, so there was no cross-country travel; and it seemed at first blush to be a straightforward warrant. There was an armed and dangerous fugitive at an apartment, and we were going to get him. Easy peasy.

I was in my usual spot. I preferred the rear-facing jump seat behind the driver in the Bearcat armored vehicle. I pulled back the bolt on my M4 carbine to check that there was a round in the chamber. I tested the light, laser, and optic even though I had just changed the batteries last night. I made the same checks with my 9-mm Glock pistol. I had tested both radios and my headset and checked that my flashbangs and medical kit were intact.

Over the back-up "beeeeep-beeeep" of the Bearcat reversing out of its spot, I thought I heard someone humming, "Dum dum dum, dudda, dum-dum." The truck stopped outside the garage and other vehicles lined up behind us. Vulcher looked out his bullet-proof window to make sure all of the vehicles in our convoy were together and in order.

The rest of the assaulters, electronics technicians, crisis negotiators, TOC personnel and the ASAC were in their own cars stacked up behind us.

Mackey's wife, a fellow Special Agent and a crisis negotiator, sat in the front passenger seat where she would have the best access to the PA system on the truck. One of our new operators, Dean, was standing on a pedestal between Vulcher and me in the passenger-side jump seat.

I first met Dean at the ambulance in Dripping Springs. He was a lean, athletic former Ranger from Second Battalion, 75th Ranger Regiment, and I thought very highly of him. Always a cool head, but aggressive in his tactics. Always ready to give a good suggestion and mature beyond his years. Dean was getting situated in the turret since he was assigned to cover the entry team from the top hatch of the vehicle should we need to approach the residence on foot. The guys in the bench seats rested their legs on and between our Avatar robot, breaching equipment, and medical supplies. The space under their seats housed our 40mm tear gas guns, extra ammunition, and extra flashbangs. It was always cramped in the truck. The Bearcat was newer and more modern than the Hogg, but it wasn't big enough to hold even half of the team. We crammed in tight.

I had my war bag hung on the strap by my head and my rifle tucked between my legs, muzzle down. My war bag had my gas mask, medical supplies, a stuffed animal for any children on scene, and the Old Glory ball cap that I wore when the helmets came off.

"Dum dum dum, dudda, dum-dum." The humming was coming from above me.

I looked up into the turret. Dean was looking out across the parking lot from his perch.

I could hear the lyrics forming in my head. *Dum dum dum, dudda, dum-dum.*

That's "Ice Ice, Baby!"

"Dude, are you humming Vanilla Ice?" I asked, laughing. "That's going to be in my head all day! Stop it!"

He peeked down from the turret and stared at me quietly. He didn't know my sense of humor yet.

"No more humming Vanilla Ice!" I smiled at him.

"It's not Vanilla Ice."

"Okay, 'Vanilla Dean'," I joked with him.

Vulcher laughed and belted out, "'Nilla Dean!" More laughter started to erupt in the truck. It was the sort of humor that was fueled by caffeine and lack of sleep. It was 5 a.m., but most of us had been up since

3 a.m. to gear prep. It was a nice cut in the tension that new operators often had before ops.

"Nilla, please!!!" someone yelled out from the back of the truck, and the laughter got louder.

"Myyyy Nilla!" someone else said in their best Denzel Washington voice.

"Nilla!" The guys erupted from the back of the truck.

That name's gonna stick. I smiled to myself. *God, I love these guys.*

Every vehicle had pulled out of the gate and lined up in the roadway outside HQ. Vulcher gave the order over the radio, "Move out." The truck lurched as it shifted into drive. That was a fun way to start the morning, but now it was gametime. The jokes were over until we got the bad guys and all had breakfast together later this morning.

"It was 'Under Pressure'." I heard from the turret.

"What?" I looked up at Nilla.

"David Bowie. 'Under Pressure.' It's not Vanilla Ice," Nilla repeated.

"Okay, Nilla." I said and gave him a thumbs up. "Stay frosty up there."

ROLLING GREEN

Once we were on the road, I keyed up my radio. "Voodoo One, this is OC-3, how copy?"

I always liked to reach out to our pilots before the action started. Since Lil Toe had stepped down as our team's STL and Vulcher had taken the STL job, Vulcher was OC-1 now, not Lil Toe. I had been Lil Toe's second in command, or OC-2. Now, with Vulcher as STL, Lil Toe and I were Blue and Gold Team Leaders, respectively. Since Lil Toe was senior to me, I gave my OC-2 patch to Lil Toe as a sign of respect, and I switched to call sign OC-3 (third in command). I had been about every number from OC-15 to OC-2, and like most guys, I was just glad to not be a "double-digit" anymore. Double-digits often got tasked with simple but important tasks like fueling up the trucks and loading ice chests. For the higher numbers, new guys often just got the call sign that was left over from another guy who had retired. It didn't matter since anything below OC-4 didn't indicate seniority. OC-15 could be senior to OC-10. Lil Toe was retiring after this op, and I'd be OC-2 again tomorrow. I'd continue to be Gold Team Leader, Sniper Team Leader, lead medic, as well as be second in command overall.

Gobbler would eventually step up as OC-3 and take over Blue Team. Gobbler had tried out for SWAT later in his career, compared to most. He was over 40 when he went through SWAT selection, which made him sort of a badass in my eyes. Since he was an experienced Agent and a former police officer with street experience, his maturity shined during selection. He got his call sign when Wylie brought Gobbler some eggs from his ranch. He made a comment about "gobbling them up" and that's all it took. Lil Toe started calling him "Gobbler."

"Good Copy, OC-3, Voodoo is on station. How me?" It was Crew Rest's voice. Big Al wasn't our pilot for this mission, but all of our pilots were equally highly skilled and air support was worth their weight in gold. This guy was good, but he once made a comment to our exhausted team, late in a mission, about how he had to land before a mission was over because he was too tired and needed "crew rest."

"Solid copy, Voodoo. We're five mikes out." I said, letting him know we'd be at the objective in five minutes. The knobby tires on the Bearcat droned a steady rhythm that was oddly relaxing. The back doors were locked open to keep the cabin at ambient temperature, and so we had an open view of all the traffic behind us. It was entertaining to see the expression on the faces of bored commuters as we passed them and they saw into the truck. *This is what a can of whoop-ass looks like, ma'am.* Eight men with loaded rifles between our legs. Most of us with eyes closed and meditating in preparation for the operation. Every once in a while, a child would wave and I'd happily wave back.

I clicked the mic button on my chest. "One minute out," I said with an intentionally calm voice. I craned my neck over my left shoulder to look out the windshield. The target building would be in view after the next turn.

"TOC, this is OC-1, we are rolling green," Vulcher said. We would often stop before Phase Line Green to attach the breaching ram to the front of the Bearcat. That way it was already in place to pop open a door and we didn't have to leave the protection of the armored truck. In this case, however, surveillance Agents at the site had let us know that there were numerous cars parked at the apartment, and there was no way we could use the Bearcat's Dong of Justice to breach the subject's door. The Electronics Technicians, the supervisors, and the TOC pulled out of the convoy and parked a block away as we pressed on to the target location.

The Bearcat groaned to a stop. It was oriented so the right side of our truck faced the subject's front door. That gave Vulcher and our negotiator front-row seats. The apartments were decent-looking compared to some of the slums we had seen. It was a double row, townhome design, as expected. Based on the blueprints, each dwelling had a bottom floor entrance and an upstairs bedroom. The bedroom windows faced toward the parking lot and were directly above the front door.

Time to get the armed and dangerous fugitive with a gnarly methamphetamine habit. Again. It all felt very routine. I had done nearly two-hundred high-risk operations at this point and had arrested over three-hundred subjects. I still loved SWAT, but each mission felt more like work and less like a thrill. Either way, the work was righteous. The road trips and time away from home was getting old – as was getting up at 3 a.m. in yet another unfamiliar hotel. Flying or driving across the country for big ops was still the best, and that's what made every bad day behind a desk worth it. We joked that missions like these – local criminals – were the best "training" we could get. It *was*, in fact, a high-risk operation, but it sometimes felt like an easy win because we stacked the odds so highly in our favor. But these "easy" ops kept us sharp for the complex national ops, like Delta Blues.

CALL OUT

The PA from the Bearcat screeched as we started the call out. Voodoo was watching us from overhead. "This is the FBI. You're surrounded. Come out with your hands up. We have a warrant."

How many states have I been deployed to for SWAT? I counted on my gloved fingertips. *Oregon, Utah, California, New Mexico, San Juan, Florida, Arkansas, Texas, Kansas, Arizona…Plus all the places I've been sent for training: Virginia, D.C., Tennessee, Massachusetts, Mississippi, Missouri, Colorado…Maybe I should count what states I haven't been to…*

I tended to want to count things when adrenaline started to flow and my concentration went up.

…Head back in the game, man. I physically shook my head to heighten my alertness. The heater was too high in the truck for my tastes, and I worried the warmth could make me complacent. My favorite temperature for an op was 40 degrees Fahrenheit. Above freezing was always good, but with all this armor on, 50 degrees would make me sweat. My preference would have been to leave the heat turned off in the truck so that when I stepped out into the winter chill, my eye protection and my optic didn't fog up from the temperature change. I tried to leave the temperature to the driver and not to complain.

Gobbler, Default, and Uptown were in the Bearcat with me. Uptown hadn't been on the team for even a year yet, but he fit in perfectly. He was my equal when it came to nerding out on ballistics minutia and enjoying sarcastic humor. His original call sign was "Pointy" from an incident where he and Crash (a.k.a. Crashendoffer) were tasked with the simple task of backing the Bearcat out of the garage. They smashed it into the top of the garage door before they realized it wasn't up high enough. The incident was caught on video and replayed for the entire office during an annual conference to display their poor judgment. During the debacle, Pointy futilely pointed to the point of impact between the garage door and the expensive armored truck. Crash got his name from the same event. You can imagine how.

Uptown didn't much care for "Pointy," and he tried futilely to adopt "Pyro" or "Backdraft" as his call sign after an op in which he threw a flashbang into a room that lit a mattress on fire. But no one was going to let him have a name that cool. Then one fateful day in the team room, he made an innocent comment about how his dad could help us with financial advice, because he owned his own firm. Vulcher called him "Uptown," and he instantly hated it. That was all we needed. The next week, I had his name tape embroidery on multicam camouflage material with a Velcro backing so he could affix it to the back of his ballistic helmet. Once it was embroidered, it was official. I sent the following message to the entire team the next week:

> Because you once mentioned a small detail of your father's financial success to a team you mistakenly thought you could trust with private information; due to your newly assumed silver-spoon existence, and despite your dirty, dumb, Marine, knuckle-dragger exterior; because Pointy doesn't carry the weight of shame it intended; and despite your efforts to burn down a house to avoid this name, we dub thee . . . Uptown.

The PA blared commands to surrender again, and while some neighbors were peeking out their windows at us, no one was coming out of the subject's apartment. Just as we rehearsed, we would make an approach to the front door if the callout failed. Even though Gobbler and I were in the truck with him, Vulcher used his radio to give the order, "Dismount. Manual breach." That way the other truck and the TOC knew what we were doing.

Default got out first and stood behind his ballistic shield. Uptown got out next, stood to Default's right, and held his shield up to the edge of Default's shield, forming a double-wide simple phalanx. Gobbler stood behind Uptown, and I stood to his left, behind Default. Gobbler and I shouldered our rifles. I covered the left windows, and Gobbler covered the right ones. There was one window on either side of the door and a double window on the second floor above the front door. I switched my rifle into my left shoulder to better shoot around Default's shield. Nilla was covering the upstairs windows from his perch in the turret of the Bearcat.

Headlight got behind me and squeezed my arm to let me know he had the breaching gear and was ready to move. Lil Toe joined us as well. Gobbler and I nodded to each other and then we simultaneously squeezed the triceps of the shield men in front of us to let them know everyone was ready.

We moved swiftly in unison to the front door. I suddenly felt fully awake and alert in the crisp winter air. The colors looked vivid, I felt strong, and air felt full and deep in my lungs. I was exactly where I was supposed to be – with my tribe.

At the doorway, we stacked to each side so we were out of the way if bullets came ripping out of the door.

"Chance Clark, this is the FBI. We have a warrant for your arrest. Open the door now." I used his name to avoid any confusion. Now that we were out of the protection of the Bearcat, time wasn't on our side anymore. Every second we stood here, we were losing initiative and exposing ourselves to gunfire. I knew from studying the apartment blueprints that he could shoot from the bedroom windows above us or from the kitchen window on our right flank. Even with two shields up, we were putting our lives at risk. One round from the upstairs window from a rifle would punch right through our helmets.

"Breach and hold," I said calmly over Fight-Net so that everyone including the TOC and teammates back in the Bearcat would know the plan. Headlight stepped in front of the team and set a hydraulic pry tool at the base of the door. He wedged the teeth of the tool between the door and the frame, then got behind the shield with his remote control in hand.

"Breaching," he said quietly. The breaching tool softly whirred in his hand, and through the cable that led to the wedge, hydraulic pressure quietly built up and spread the teeth of the wedge until it warped the door right out of the frame – deadbolt and all. The door softly and slowly started to swing open with just a gentle nudge from Headlight's boot. After opening only three inches, it got hung up on the splintered door frame.

With a swift kick, Headlight popped the door the rest of the way open, and we were met with the sight of a man racing downstairs, buck naked, holding a metal bar we could only assume was for jamming the door.

We all froze as his eyes met ours. It took him a second to realize his timing to bar the door was off by a few critical seconds. After an awkward millisecond, the man spun around, dropped onto all fours and monkey-crawled in his naked glory back upstairs.

"Was that him?" I asked, stunned.

"I dunno," said Uptown. "I only saw balls and butthole."

Whoever it was, he heard us calling the whole time and had been using that time to plan to bar the door until we interrupted him.

"This is the FBI. Come down here right now. Do not make us come up there to get you!" I said in a low and somber voice. From upstairs, I heard a man's voice answer calmly, almost singing, "Oh. Okay. I'll be right down. Just a minute." Then eerie silence. All the hairs stood up on the back of my neck. Being just outside the house right now was worse than being inside with the bad guy. I didn't want to stand exposed in the porch doorway a second longer. For all I knew, creepy-naked-guy had run upstairs to retrieve his rifle and now he could shoot down from his upstairs bedroom windows and kill us all.

"Bang and clear," I called.

In less than a second, Lil Toe hucked a bang into the entryway, and it detonated in a fantastic fireball. The volume of smoke from this particular flashbang caught me and Gobbler off guard. There must have been a decade of dust and dander on that carpet, because it blew up a burning smoke cloud that made it look like the air itself caught fire. I'd never seen anything quite like that.

The pungent odor of burnt hair and scorched carpet filled my nostrils. We couldn't see inside the structure at all now, due to the smoke, and our backs were exposed from the top window. Moving through the smoke cloud was better than standing on the porch, so I stepped deliberately into the house, knowing that my teammates would follow.

As I entered, I snapped my rifle to the right, where the kitchen was supposed to be, and was relieved to see a refrigerator through the smoke as a reference point that I was indeed in the kitchen. I held my position and called out, "Kitchen clear." Gobbler had already crisscrossed left and covered up the stairs.

Gobbler and Lil Toe stayed on the stairs while Uptown, Default, and I combined to clear the front room. Suddenly, a woman popped out at the top of the stairs. Oddly, she was brushing her teeth.

"What's happening? He's not here!" she said.

Lil Toe ordered her to come downstairs, placed her in handcuffs, and walked her out to the holding area.

"One in custody. 'No joy' on the suspect," I called on the radio. "Downstairs clear." I looked at Gobbler. "I don't feel like chasing a crazy naked guy upstairs," I said.

"Me, neither," said Gobbler "Let's get Cletus."

Lil Toe grabbed the Avatar robot from the Bearcat and put it at the base of the stairs. Gobbler kept his rifle pointed up the stairs while the robot started to climb with its articulated tracks. If the subject wanted to pop out now and shoot at us, it would go poorly for him.

GAS IT

In a few minutes, Cletus successfully crawled up the stairs and cleared the upper floor of the townhouse. Our fugitive was nowhere to be seen. If he was hiding under the bed or in attic space, we were going to tip the odds in our favor before entering. We had no intention of making this a fair fight. If you fight fair, your tactics suck.

Vulcher keyed up his radio. "Gas it."

"Copy," I replied. Lil Toe sent most of the team back to the Bearcat to mask up while Default and Headlight held their positions at the entrance. I made my way back to the Bearcat and climbed into the back, opened the access panel under the bench seats, and pulled out a Heckler and Koch 40mm grenade launcher. The rest of the team reformed at the entrance to the house to hold that position while we let the gas do its work. I cracked open a can of gas grenades and slid them to the back of the truck. After I opened the shooting port on the right-side rear door, I positioned myself so that I could shoot directly at the upstairs windows and stay completely behind the armored back door.

"Blue team standing by," Lil Toe called on the radio.

"Gas out!" I called as I loaded a canister of OC (oleoresin capsicum, a.k.a. pepper spray) into the launcher and cocked the hammer. It looked like about a 50-yard shot. The canisters were big, heavy and slow, so you needed to know your range to lob them into a window.

"Send it," Lil Toe called.

I pulled the trigger on the grenade launcher. *BLOOP!* The gas launcher kicked like a 12-gauge shotgun. The canister smashed through the left upstairs window and screen and ejected a purple mist of pepper spray into the room. No sooner had it fired than I cracked the breach-loading gas gun open. I pulled out the spent casing, and I popped in another round. I put four rounds each in the two adjacent window panes in quick succession. On the last round, I pulled the shot at nicked the bottom of the window sill. Some of the angry pepper juice splatted down onto the front door. I could see Default shaking his head and coughing as he stood in the entryway on the first floor. Little drops of purple OC dripped around him. *I'm gonna catch hell for this.* The OC was also building up in the apartment and leaking out the front door. The half of the team that hadn't put on their masks yet was feeling the effects of the pepper spray mist.

Vulcher's voice crackled on the radio, "Clear the second floor." OC wouldn't hurt anyone, but hopefully it would slow any bad guys down if they were waiting to ambush us. Our aircraft had been orbiting overhead the entire time, scanning the area with thermal sensors. There was no sign of escape, so he had to be in the building, waiting for us upstairs.

Blue team started up the stairs and cleared through the structure. The guys who hadn't masked up yet joined in, knowing they were about to get a wallop of pepper spray. I would have done the same thing. I hated to be outside looking in. It felt like an eternity, waiting to hear them call that they had arrested the fugitive. I set the gas launcher down and pulled my rifle from behind my back. I wanted badly to be in the building with my teammates.

My radio clicked in my headset. "All clear upstairs. Starting secondary searches." Lil Toe's voice sounded raspy. He had decided to lead Blue team upstairs, although he hadn't had time to put his mask on. Nilla and Headlight seemed immune to the OC, but Default and Gobbler came out the front door and looked terrible. Default didn't have his mask on. Gobbler did, but he must have had a leak because his eyes were red and snot was running down his face. I called them over to the Bearcat, switched hats, and became their medic.

"There's attic access on the second floor of the apartment." Lil Toe said over the radio.

"Come sit down, I'll get eye wash for you guys," I called to Gobbler and Default.

Gobbler set his mask on the floor of the Bearcat while I grabbed my med bag.

"Dude, where's your gas filter?" I asked Gobbler.

"What?" Snot was dripping down his face like a waterfall. "Oh, crap." He realized that his gas mask filter wasn't attached. We had two filters to choose from. The smaller, flush-fitting filter was for law-enforcement work like tear gas. We also had a larger filter that was rated for WMD missions and could filter out lethal agents like anthrax or ricin. Our gas mask bags weren't big enough to store the mask with the larger WMD filter already attached. So, you had the option of putting it in the bag with the smaller filter already attached, or just place the bigger filter in the bag next to your mask. Gobbler had chosen the latter, and in his haste, forgot to attach the filter.

Vulcher was in his jump seat, and I was in the back of the truck when Lil Toe walked over to us. "He has to be in the attic. We can grab a throw-bot and hold it in the attic access and use it like a periscope," Lil Toe said.

"That's a good idea, but if he sees that robot come through the opening, he could shoot down through the ceiling. Let's gas the attic," I said.

"How do we do that?" Vulcher asked me.

"I'll shoot through the second-floor ceiling from the bottom of the staircase."

"I like it. Money, you take a team and do it," Vulcher said.

"Roger that," I said.

"Default, are you good to go?" He nodded. "I need you on shield. Cover me. We're going back in to gas the attic. Headlight grabbed his rifle and joined us. I needed someone to cover me while I was launching gas. I slung my rifle behind my back, threw six gas canisters into the dump pouch on my belt and grabbed the gas gun.

Once we got to the apartment, we walked through the open door. Default held his shield to protect me from anyone who might shoot down the stairs. I cocked the gas gun and aimed for the ceiling at the top of the staircase. *BLOOP!* The launcher fired and the gas canister punched through the second-floor ceiling and misted its purple pepper spray into the attic.

From the bottom of the staircase, I shot four more gas canisters through the ceiling into the attic. If he was in the attic, he was naked, cold, and miserable. We could slow down now. He had to come out at some point. Honestly, we could just order pizza and camp out in the parking lot all day. There was no escape for this guy.

Meanwhile, Needle-D and Nilla went across the parking lot to the mirror-image apartments and determined if the attics connected – and they did except for an occasional firewall. Now we had to consider if he had "Jason Bourne'd" himself from one attic to an adjacent attic and entered an apartment on the right or left. For the safety of the inhabitants on either side, Vulcher decided that we would "knock and talk" with the neighbors and ask if they would consider evacuating on account of a meth-smoking, naked crazy dude who might come climbing out of their attic access. We were at the end of a row of twelve contiguous townhomes. The one we just cleared was #11. We had to consider that our subject had made access to #10 or #12.

"Money, you take Gold to apartment #10. Lil Toe, you take Blue to the right to #12." Vulcher ordered.

I looked at Nilla. "Hey, do these all connect? Could he be in any of these apartments on the left?" Nilla shook his head, "The apartment manager said there is a firewall between every four apartments. Needle-D and I crawled up into the attic of the apartments behind us. If they are the same design, there is definitely a firewall for every four units. From the attic, he could only get into nine, ten, eleven, or twelve."

We cleared apartment #10 while Blue team waited. It was best to let one team work at a time, so we could back each other up. Apartment #10 was empty, so we held our position while Blue team cleared #12. With the last three apartments on the left of the structure clear, the final one he could be hiding in was #9. It was logical to clear #9, but I had my doubts he was still here. We had been out here a while. I hoped he hadn't snuck out and evaded the perimeter team somehow.

I called over my radio for Gold team to rally up on me and told them the plan to knock-and-talk and search the last apartment. In the meantime, Blue team was going to search the subject's apartment (#11) again. Before we went to apartment #9, a neighbor waved me over, so I decided to chat with her.

"How's it going, ma'am?" I asked.

"It's an exciting day for us." She laughed nervously, and I smiled.

"I assume you're looking for someone," she said. "Listen, most of these apartments are empty. But my neighbor told me she had to lock her attic door because someone had gotten into her apartment that way before," she said.

"That is very helpful, ma'am." I appreciate it. With a nod, I excused myself and walked with Gold team to apartment #9. We stood out of view of the windows because of the small possibility that the fugitive was inside and planned to shoot us. We needed to be safe, but I also didn't want to terrorize an innocent neighbor in the middle of the day. I knocked loudly on the door. If the person who came to the door was relaxed, they probably weren't being held hostage by a crazy naked guy. If they seemed scared, I needed to ask permission to clear their home.

No one answered the door. I knocked again. I hated to damage the apartment, but I didn't have much choice. It was either unoccupied or there was a fugitive hiding inside – possibly using the rightful tenants as hostages. Either way, we had to get inside.

"Breach and hold," I said to Headlight. He set the hydraulic tool on the left side of the door while I stood to the right on the hinge side. He made eye contact with me and I nodded. He pressed the button on his remote and the tool whirred quietly until the door frame groaned and eventually cracked. A sliver of silver emerged from the door as the deadbolt slid completely out of the frame. But it didn't open. Headlight gave it a mighty kick, and the door only opened a couple inches.

"It's blocked with a table or something." Headlight said.

Suddenly, through the tiny gap, I saw the figure of a man standing on the stairs to the left of the door. The man was dressed in blue jeans and was bald. He seemed familiar, but I felt inexplicably alarmed. He was oddly calm. If anyone should be alarmed, it should be this guy. Men in camouflage uniforms and carrying rifles just broke his door frame.

"Hey, I'm getting ready for work," he said. It was almost the same sing-song voice as the naked subject we saw next door. The hair stood up on the back of my neck. I couldn't place why I was spun up, but I could feel adrenaline surging. I barked out an order before I really thought too much about word choice.

"Get your hands up, now!" I yelled.

"I'll be just a sec." *His voice is too calm. Something's wrong.* He turned and started walking up the stairs. I couldn't see his hands and I couldn't discount that he was armed or going to arm himself. I couldn't afford to lose sight of him. I smashed my body against the door, but it didn't budge. There was an odd black outline around the top and sides of the door. I threw my shoulder into the door again. It still wouldn't open more than a few inches.

"Show me your hands right fucking now!" *That was a bit much. He's probably going to file a complaint. Why am I so spun up?*

He came back down the stairs and put one arm out the door. "I can't get out. There's not enough room," he said, almost whining.

In that second, Default latched onto his arm and pulled it with all his might. Astonishingly, he pulled the man completely through the small opening in the door. It was like seeing someone pull a 200-pound tub of Play-Doh through a press. I was relieved to see Default was as alarmed as I was. Once the man was out of the door, he started swinging at Default, but didn't connect. Default and Gobbler tried to take hold of him, but he continued to try to fight and get away. The three of them were quickly on the ground in a short-lived wrestling match. The man was screaming like a child. "What?! I was just working in there, man!"

I held my rifle up to cover the entrance and resisted the urge to join in on the fight. Default put the bald guy in an arm bar, while Gobbler put handcuffs on him. Once he settled down, I saw several odd patches of hair on his otherwise bald head. It was like it was poorly and hastily shaved.

"You got a shitty haircut, Chance." Headlight said. He must have seen the patches of dark hair on the back of his head, too. The fugitive's shoulders slumped when he heard Headlight call him by his name. The gig was up, but what a spectacular performance.

Gobbler walked the subject to the back of the Hogg and turned him over to Agents on the perimeter. In the meantime, Headlight and I tried to push the door open the rest of the way, but it wasn't going to budge.

I keyed up my radio and looked across the parking lot at Vulcher. "Subject in custody. Gold team has a failed breach. Alternate breach, alternate breach." Vulcher had seen everything unfolding from his position by the Hogg.

"Blue team, link up with Gold and execute an alternate breach." Vulcher announced on the radio.

K2, South Beach, and other guys joined up with us as we moved to the back of the apartment. We grabbed a ladder and scaled the back fence.

When we got into the backyard, K2 breached the backdoor by busting out a sliding glass door. We moved methodically through the apartment to clear for other threats. As we reached the front of the apartment, I could see that the black outline around the front door was actually a black refrigerator that the fugitive had used to barricade the door. *That's why it wouldn't open all the way.*

We cleared upstairs and found there was a plywood-patched hole in the closet that led to the next apartment. Someone had once punched through the drywall, allowing access between apartments, and it had been repaired with plywood to prevent another attempt. More than likely, he had been planning to enter this apartment, as he had in the past, and found the plywood blocking his path.

In the upstairs bedroom, there was a pillow jammed into a hole in the ceiling. When we pulled it down, a cat scratching-post fell out with it. Unpacking this crazy dude's wild journey was almost as entertaining as watching him being pulled through the door like taffy.

"So...he smashed through the ceiling from the attic," I said. He must have made an impromptu entrance into the apartment via the ceiling after he hit the plywood roadblock in the closet where he had come through in days past, as the neighbor had described.

Default chimed in. "Then he crawled back up there and plugged the hole with a pillow to try to keep the gas from getting in." All of our eyes were watering and red from the gas. This was just residual gas from the apartment I had pumped OC into two doors down. The gas must have been running through the attic that connected the four units after he fell through the ceiling, and he engineered a way to slow the flow of gas by jamming a scratching post with a pillow in the hole. In the bathroom, there was a shaver and a pile of our subject's hair in the sink.

This operation had been named Operation Vatos Locos, or "crazy dudes," several months ago by a hard-working Case Agent. I don't know why he picked that name. He could have named it Purple Rain, Lucky Charms, or Blynd Justice – just a code name like we often saw. I'm certain this Case Agent and his squad mates had spent countless long nights piecing together surveillance records, financial transactions, pen registers, and undercover operations. Finally, with three dozen search warrants and arrest warrants in hand, we got to start dismantling this broad-reaching criminal enterprise. Today's arrest made the operation code name apropos. This was definitely a "crazy dude."

During the secondary clear, I checked the sliding glass door that K2 had shattered and saw that it was unlocked.

"Hey, K2?" I said loudly enough for everyone to hear. "At breacher school, do they teach you to check to see if the door is locked before you smash it in?" I smiled at him.

"Wouldn't you have been more useful sleeping in the woods?" He shot back and made me laugh.

Back at HQ we parked the Bearcat outside the garage and started cleaning out the vehicle. There were expended 40mm gas canisters and water bottles on the floor, as well as extraneous gear that needed to be reclaimed. As we were finishing up, I saw an extra gas mask on the floor of the truck and said, "Hey, guys. Whose mask is this?"

Vulcher asked, "Does it have a filter on it?"

"Yep," I said.

"Then it's definitely not Gobbler's!"

Chapter 14
Whom Shall I Send?

Around 2003, Wylie and I were brainstorming about the creation of a Challenge Coin for the Oklahoma FBI SWAT team. He knew that my mother had taught Latin at a university, and we asked her to translate Isaiah 6:8 into Latin to use as our motto. The translation is "Hic sum, me mitte." From that day on, the signature line on my *fbi.gov* email account included that phrase. It has been used by a lot of people in a lot of circumstances, which isn't surprising for a scripture that was written 2,700 years ago.

Around 2007, the guys at the SWAT Operations Unit at Quantico standardized the national FBI SWAT motto as the variant, "Mitte Me." For me, *mitte me* means someone has to be in the trenches. Despite the million moving parts in a battlefield, someone must choose to step into the mire.

During twenty-one years on the team, I learned how to breach a pressurized aircraft, drive a locomotive, fly an airplane, rescue a hostage, disarm a nuke, treat a gunshot wound, and how to hit a target a half a mile away. But the important thing was that I learned servant-leadership. I learned how to follow willingly and how to lead with humility. Everyone had earned a place on the team, yet each team member put himself last.

During SWAT selection trials, it was important to see how well the candidates fired their weapons, how they responded under confusion and stress, and how physically fit they were. But the pivotal moments were times when we tested their moral code. Sometimes we would put out a case of twelve water bottles for fourteen candidates and watch what unfolded. If you were the type of guy who would rush to get a bottle first and walk away, it didn't matter how fit or smart you were, you were not cut from our cloth. The guy that would count them first and recommend a way to share twelve bottles among fourteen men already had what it took to be an operator. He might not have had the best time on the run, but he already had the roots to join a team that was based in respect and mutual submission.

Despite what popular movies and television shows would have you believe about the FBI SWAT team, it is a fellowship primarily consisting of devoted husbands, fathers, and men of faith. That's what I loved most about the team. It was an environment where you had the freedom to speak what you believed about politics, church, marriage, or any topic that's not usually acceptable in large groups. There was grace.

You could tell a bad joke, or have a beer after a mission, but you could still ask for prayer from your teammates when a family member was sick or you were struggling in your marriage. We often prayed for each other on and off duty. Guys knew that I loved ballistics and aviation, but they also knew that Jesus Christ is the absolute center of my life and wouldn't hesitate to ask me to pray for them. I was blessed to belong to a group of people who would not only protect my back in a gunfight, but that I could trust to share all my thoughts and concerns with as I try to unpack my purpose on this planet before God brings me home.

You got to meet some of my brothers who have sworn an oath to protect you and the Constitution. We don't swear to obey any politician or manager; we only swear to obey the Constitution of the United States of America. We are former Deputy U.S. Marshals, Rangers, scientists, SEALs, CIA, NSA, Air Marshals, Marines, policemen, accountants, attorneys, family men, patriots. Men who embody our motto: Fidelity, Bravery, Integrity. Men who love to laugh over a good story, learn a new grappling technique, play chase with their kids, lift weights, read a good book, and hike in the mountains. Athletes, detectives, scholars, and gentlemen – each and every one. Men who conduct operations to arrest criminals in your neighborhood as well as overseas, who train for unthinkable missions, some of which you may not have heard about until now. These men are, each one, ready for the fight. That's my team. My brothers.

Yet, this hasn't been just stories about a team. And it hasn't been about psycho drug dealers. It's about little girls locked in closets. It's not about the cartels, but about the child that grows up in a fifth wheel behind a drug dealer's safe house. It's not about the lady that makes ricin toxin; it's about her children, and an unborn child who almost died from being exposed to poison. It's not about a bomb maker at the base of a mesa in Oklahoma; it's about the people he tried to murder in the church. It's about the terrified children who had their tears wiped away by camouflaged men who came in the night. Hostages rescued. Homeowners freed from the harassment from drug dealers next door. We saw the ugly underbelly of humanity, pulled on our boots, and set out to make things right again.

We came together to pick up the pieces, love the victims, prove the facts, and ensure justice – despite the risk to our own lives and the time away from family. It was painful some days, but the journey was a joy, and the mission is righteous. I always had the team to laugh and cry with when the late nights were done. They are the best this nation has to offer, and they are the greatest of friends. I'm honored to have had each moment of honest criticism, each word of encouragement, each handshake and hug in the hallway.

I couldn't have asked for more.

Now you know, at least a little bit, about the men and the mission that I love so much. Honestly, there are a hundred more stories, just not enough time to write them.

There are dozens of ingredients that made us a great team, though – and those, I can write.

>Driving 100 mph on an abandoned runway.
>Riding choppers five feet off the water of the Cimarron River.
>Treating a subject's gunshot wound.
>Treating an operator's gunshot wound.
>Watching a man die – a man who almost killed you.
>An uncontrollable laugh over beers in San Antonio.
>A poker game with your rifle slung on your back in a hidden room at the Super Bowl.
>Spontaneous foot races and unfortunate gutter fishing trips.
>Arresting an entire small town in Arkansas.
>A night maritime infil to a bomber's remote cabin.
>Busting a water main with an MRAP.
>Water survival tests that look a lot like pool parties.
>Taking down the Zeta cartel money-laundering horse ranch in Lexington.
>Banging an empty apartment because the curtain moved. (Probably because a fan blew it.)

Hours sweating in a van waiting for a child molester to show up at the Post Office so we could arrest him.

The entire team taking turns peeing in a tiny church bathroom in full assault kit.

110 degrees in the Tulsa woods waiting for an extortion subject to pick up a million in cash.

Patrolling into a cockfighting ring in Arkansas full of corrupt cops.

Rescuing a hostage in Carney. One dead hostage-taker.

A llama that looks a whole lot like a man in a white t-shirt.

Protecting Comey, Mueller, Freeh, Wray, Ashcroft, Reno, Lynch, Sessions, Barr and Holder…or at least being prepared to burn down whoever tries to kill them.

Kicking in the door to Moussaoui's dorm room on 9/11/01.

Searching the pockets of Heinrich's corpse while the bomb blast is still smoldering at the University of Oklahoma.

Stand-by while the Oklahoma City bomber breathes his last evil breath.

Swim test. Fitness test. Firearms quals. Trauma care. Rappel. Patrol. Grapple. CQB. WMD. Tubular assault. DigPro. Lather. Rinse. Repeat.

Ten thousand sim rounds.

Five hundred thousand live rounds.

A thousand bangs.

A hundred gallons of sweat.

A million laughs.

The ingredients of a good team are hard to explain to someone who hasn't been one of us. And it's hard to sum up over two decades of being part of a team like ours. But for me, Isaiah 6:8 (ESV) is as close as it gets:

"And I heard the voice of the Lord saying, 'Whom shall I send, and who will go for us?' Then I said, 'Here I am…'"

Mitte me.

Send me.

ACKNOWLEGDMENTS

To my editors:
Justice Rebmann and Judith Farrar

To the Team:

Vulcher, a.k.a. The Situation
Gobbler
Roscoe
Headlight, aka Mugshot
Bourdain, a.k.a. DrillBit,
a.k.a. Mr. Slave, a.k.a. Sharkbite
Pointy, a.k.a. Uptown
Crashindoffer
Red, a.k.a. Flounder,
a.k.a. Sailor Jerry
MacGruber
South Beach
Default
K2
Nilla
Ruts, a.k.a. Ryan
Other Ryan
Third Ryan, a.k.a., R3
Grumpy
"Tim"
Scuba Steve
Needle-D
Dangler, a.k.a. Gerlad
Lil Toe
Twilight

Mackey
A-Train
Big Toe
The Major
Bodie, a.k.a. "Lieutentant" Dan
Frosty
D.B.
JimBo
J.J.
Z
Shallow, a.k.a. Gene-Pool,
a.k.a. Mean Gene
Mag
Wylie
Febreze, a.k.a. Monkey Doctor,
a.k.a. Monkey D
Q Dawg
"Billy White"
Nestle
Lobo
Grapenuts
Radio
Big Al
…And all the others.

ACKNOWLEGDMENTS
(Continued)

To Lt. Col. (Ret.) Dave Grossman:
Thank you for your encouragement, friendship, and for your work, "On Killing." You helped me understand that it's not absurd for a Sheepdog to love his Sheep, love the Great Shepherd, to be disgusted by killing, but to still be eager to kill a Wolf.

To my old friends and former colleagues:
Many thanks to the FBI case Agents, teammates, pilots, analysts, professional support; the Assistant United States Attorneys; and agents and officers from sister agencies to who reviewed the facts in this book for accuracy. Working with you was the honor of my life.

To my book cover artists:
Matt Gambrell
David Justis

ABOUT THE AUTHOR

Jeremy D.O. Rebmann

Jeremy served 32 years in the U.S. Air Force and in the Federal Bureau of Investigation. While in the FBI, Jeremy spent 21 years as an FBI SWAT operator. He served as a team leader, firearms instructor, lead medic, and lead sniper.

www.fbisniper.com

GLOSSARY

AFSOC: US Air Force Special Operations Command located at Hurlburt Field, Florida.

ANFO: Ammonium Nitrate and Fuel Oil. A type of homemade bomb.

AOR: Area of Responsibility.

ASAC: Assistant Special Agent in Charge. A middle manager at the GS-15 pay grade.

Assault Rifle: A medium-sized semi-automatic rifle chambered in an intermediate caliber, like an M16 or AK47. Typically, larger than a SMG (MP5) and smaller than a battle rifle (SR-15).

Battle rifle: A large semi-automatic rifle in a large caliber (AR-10, SR-25, M14).

BDU: Battle Dress Uniform. A style of military uniform common in the 1990s to 2000s.

Bearcat: A modern armored vehicle.

Breacher: A big dumb knuckle dragger SWAT operator who's not smart enough to graduate from sniper school.

Brevity Code: A shortened term used to quickly describe a previously well-defined object. ("The low ground southeast of the objective that is south of the main road and north of the dock will be referred to with Brevity Code 'Corvette.'" "Where is the sniper team?" "They are at Corvette.")

Bu-Car: A vehicle owned by the FBI and used for official business. Pronounced BYEW-Car.

Carbine: A short-barreled, handy rifle.

Case Agent: A Special Agent of the FBI who manages, oversees, and executes a specific federal investigation. Usually of the GS-10 to GS-13 pay grade. A term usually used for larger cases that require more than one Agent assigned. Not to be confused with Lead Agent. ("Who is the Case Agent on this? I want to talk to him to make sure I shag this lead right.")

CBRNE: Chemical, Biological, Radiological, Nuclear, and Explosive.

CIRG: Critical Incident Response Group. A national entity within the FBI that manages SWAT, HRT, HazMat, surveillance and aviation assets to respond to national emergencies.

COA: Course of Action. The specific plans for a team to execute on target for a specific situation.

CQB: Close Quarters Battle.

Crye Combat Uniform: A modern combat uniform with integrated knee and elbow pads.

Dope: The shooting data input into a scope for a precision rifle derived from ballistic data (i.e., humidity, elevation, barometric pressure, temperature, ballistic coefficient) to determine hold-overs, wind-hold, and off-angles that allows a marksman to make a first shot hit in real-world conditions. Often incorrectly referred to as D.O.P.E, a recent reverse-acronym for Data On Previous Engagements.

Hummer: a.k.a. HMMWV (Highly Mobile Multi-Wheeled Vehicle), or "Humvee".

Defilade: A place where bad guys can't see you to shoot you.

DIGPRO: Dignitary Protection. Specifically in the FBI, providing protective service for the FBI Director and the U.S. Attorney General.

Doff: Fancy military term for taking stuff off.

Don: Fancy military word for putting stuff on.

Dry Hole: A location where a subject was expected to be located, but was not found. ("He's not here. Dry hole. Let's move to the next location.")

ET: Electronic Technician. A professional support position in the FBI that provides radio and CCTV expertise to include cryptologic encoding of radios, and maintaining radios and radio towers.

FBI National Academy (The "NA"): an 8-week academy hosted by the FBI at the FBI Academy grounds in Quantico, VA to provide leadership training to senior police officers. Often a requirement to become chief of police in many departments. NA attendees share dorm space, chow hall, and classroom with NAT's but are not being trained to be FBI Special Agents.

FBI New Agent Training: 5-month-long course of training to become an FBI Special Agent. Held at the FBI Academy in Quantico, VA. Not to be confused with the FBI National Academy which is co-located at the FBI Academy grounds at Quantico.

Fight-Net: The frequency used only by operators during the high-risk phase of warrant execution. Its use is limited to operators and TOC personnel to prevent distractions from mundane questions from Command Post, negotiators, or supervisors during critical stages of high-risk operations. Typically, on a simplex frequency that is optimal for short range communication.

Flash-bang: A hand-grenade-sized pyrotechnic diversionary device that uses burning magnesium to cause a bright light, concussion, and blast that typically stuns the subject enough to diminish his reaction time.

Good-Time Radio: The radio you use to listen to music in your assigned FBI vehicle. Not to be confused with "radio" which is what is used to communicate to other Agents and the communication center at HQ.

HazMat: Hazardous Materials.

HLZ: Helicopter Landing Zone. Any area pre-designated as a safe place for a helicopter to land. ("The tennis court north of the objective is HLZ Alpha One.")

HRT: The FBI's Hostage Rescue Team located at Quantico, VA.

IED: Improvised Explosive Device.

IR: Infrared. An invisible portion of the light spectrum that can be seen with NODs and thermal sensors.

Jocking Up (v.): To don the appropriate tactical attire, e.g., armor, helmet, gloves, etc.

JSLIST: Joint Service Lightweight Integrated Suit Technology. A chemical warfare uniform.

Kit: Referring to the gear necessary for fighting. Typically including helmet, body armor, load bearing belt, as well as medical supplies, radio, ammunition, and tools.

Lead Agent: A Special Agent of the FBI that is assigned to complete investigative leads to further an investigation. Informally subordinate to a Case Agent.

Lead: A logical investigative step. ("I have three outstanding leads on this case. I still need to serve a subpoena, request a surveillance team, and wiretap.")

M4: The military designation for a 5.56-mm carbine based on the Colt AR-15 design. A short version of an M16 assault rifle.

Mike: Shorthand for a minute. ("We're five mikes out.")

Mic: Shorthand for microphone

Mobile Command Post: A command center for FBI crisis sites. Often a large RV with its own generators and radios. Typically maintained by ETs.

MRAP: Mine Resistant Ambush Protected vehicle. A genre of military vehicle that came into use to protect US soldiers from roadside bombs. Typically, much larger than a Hummer.

NAT: New Agent Trainee. A person who is attending the five-month-long FBI Academy who upon graduation is sent into the field as an FBI Special Agent. Never referred to as a "NAT." Always referred to as "New Agent." Typically meaning, one who is attending and has not yet graduated (or very recently graduated) from the FBI Academy and has not yet been given an FBI badge and credentials.

NOD: Night Observation Device.

NVG: Night Vision Goggle, same as NOD.

Objective: An area or structure that is the focal point for a tactical mission.

OSI: The Office of Special Investigations is a component of the US Air Force that employs officers, NCO's and civilians as Special Agents to conduct criminal and counterintelligence investigations.

PAPR: Powered Air Purifying Respirator. A respirator used by tactical teams to filter out poisons and allow the operator to survive in a WMD environment.

PEQ-15: A rifle-mounted aiming device that uses infrared (IR) and visible lasers.

PVS-22/26: A night vision tool installed in front of a standard rifle scope to intensify light to allow for target identification at night.

Quantico: The term FBI Agents use to refer to the FBI Academy which is located inside Marine Corps Base Quantico.

Rotor: A defunct administrative position in the FBI. A support staff member who maintained paper copies of case files in a rotary cabinet.

SAC: Special Agent in Charge. The chief executive of an FBI division at the SES pay grade (above GS-15).

SERE Training: Survival, Evasion, Rescue, and Escape. A US rigorous three-week military training school intended for soldiers or aircrew that might be forced to survive alone after crashing, ejecting, or being captured by enemy combatants.

Shag a lead (v.): To complete an investigative lead. ("Are you shagging a lead?" "Yeah, I need to pick up bank records for this subpoena.")

SMG: Submachine gun. A long gun chambered in a pistol caliber. (e.g. MP-5)

SSA: Supervisory Special Agent. A first-line supervisor at the paygrade of GS-14.

Stack: A single-file formation of operators. Typically used at a breach point.

STL: Senior Team Leader. A Special Agent at the pay grade of GS-13 that leads an entire divisional FBI SWAT Team. STL's wear designator patches and use call signs that include their division two-letter code and the numeral "1." For example, the Dallas STL would be known as DL-1.

Subject: The subject of the investigation. A person suspected of committing a crime.

Tally Ho: A brevity code to indicate, "I see the thing you are talking about."

Team Leader: A Special Agent at the pay grade GS-13 that leads one of two or more sub-components of a divisional FBI SWAT Team. Sub-teams are often color coded (Gold, Blue, Silver). On some teams, TLs will wear a designator to indicate their place in the chain of command. (HO-2 is the Houston FBI SWAT team's second in command.). TL's may also lead sniper teams.

TOC: Tactical Operations Center. A mobile Command Post this is staffed with FBI Special Agents and professional support staff that activates near a crisis site. The TOC specifically supports FBI SWAT operators via radio and telephone communications, as well as coordination with local police, fire, EMT, and hospitals. Often operated from a mobile trailer adjacent to the Mobile Command Post.

TOC-Net: The frequency used to communicate between SWAT operators, TOC, Command Post, and FBI fixed-wing aircraft and helicopters. Typically, on a duplex frequency that hits a repeater tower to allow for longer range communications. Less than optimal for direct communication between two people in the same building. Usually only used by SWAT team leaders, medics, and snipers. Modern communications headsets allowed the operator to hear TOC in one ear and Fight-Net in the other, to help keep track of simultaneous communications.

UC: Undercover Agent or Undercover Employee.

VBIED: Vehicle-Borne Improvised Explosive Device. A car with a homemade bomb in it.

WARNO: Warning Order. A preliminary notice of an order or action to follow.

WMD: Weapon of Mass Destruction

www.ingramcontent.com/pod-product-compliance
Lightning Source LLC
Chambersburg PA
CBHW050900160426
43194CB00011B/2229